GEOLOGY
UNDERFOOT
IN WESTERN WASHINGTON

DAVE TUCKER

2015
Mountain Press Publishing Company
Missoula, Montana

IS A REGISTERED TRADEMARK OF
MOUNTAIN PRESS PUBLISHING COMPANY

Library of Congress Cataloging-in-Publication Data

Tucker, David S. (David Samuel)
Geology underfoot in western Washington / by Dave Tucker.
    pages cm
Includes bibliographical references and index.
Summary: Geology of Washington State, including tours of various state
locations identified by the local geology.
ISBN 978-0-87842-640-9 (pbk. : alk. paper)
1. Geology—Washington (State), Western. 2. Washington (State), Western—
Guidebooks. I. Title.
QE175.T83 2015
557.97'6—dc23
                              2015003228

PRINTED IN HONG KONG

PUBLISHING COMPANY
P.O. Box 2399 · Missoula, MT 59806 · 406-728-1900
800-234-5408 · info@mtnpress.com
www.mountain-press.com

*Dedicated to the memories of
Drs. Maury Schwartz and Chris Suczek,
Western Washington University professors
who reviewed portions of this book
but did not live to see it in print.*

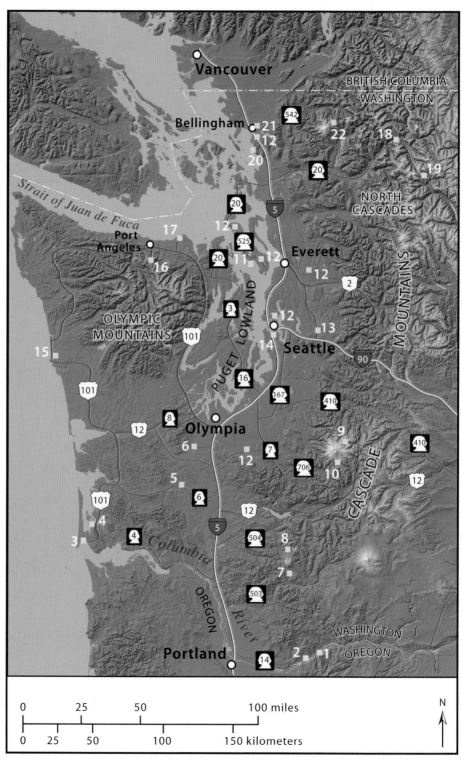

Places visited in this book, with principal highways.

CONTENTS

PREFACE

And some rin up hill and down

dale, knapping the chucky

stanes to pieces wi' hammers,

like sae mony road makers run daft.

They say it is to see how the warld was made.

—SIR WALTER SCOTT, ST. RONAN'S WELL (1824)

The region covered in this book is home to a large population of people with a keen interest in natural history. This book is dedicated to that audience. A formal education in geology is not necessary to enjoy the geologic field trips I present herein, only curiosity and a willingness to understand new concepts.

Western Washington is a dynamic geologic environment with rich, complex, and diverse geology. With active volcanoes, sediment-laden rivers, glaciers, wave-washed beaches, ancient and active faults, and a wide variety of rock types, it is a great place to experience the story of the Earth's ongoing evolution. This book spans some 90 million years of time. The subduction zone beneath western Washington is an underlying and unifying theme. If you aren't familiar with the basics of plate tectonic theory, please read that section in the introduction.

Complex and often puzzling geology generates different interpretations by different researchers. This is the nature of scientific inquiry. I presented some of the uncertainties that give rise to such controversies. There will be new interpretations as future generations of geologists develop new tools and new ideas to explain this complexity.

I am a native of western Washington. I have been backpacking, camping, and mountaineering in this beautiful area my entire life. An interest in geology is only natural in a place like this, with volcanoes rising on the horizon, the sea crashing on the coast, the evidence for recent glaciation everywhere in the lowlands, and occasional earthquakes shaking things up. I took all the geology classes I could as an undergraduate. Thirty years later, I retired from years of odd jobs, including guiding climbers up the signature volcanoes and icy spires of the Cascades, and returned to academia long enough to earn a master's degree in geology. My graduate work focused on mapping previously unstudied volcanic rocks near Mount Baker. I enjoy teaching classes, leading field trips, and giving presentations to the general public on geologic themes.

The hardest part about writing this book was deciding which places had to be left out. I included sites that are easily accessible and provide a sampling of some of the more salient features of western Washington's geology. Entire guidebooks could be written about some of the following topics. Some of the locations are briefly described in earlier guidebooks, both popular and technical. See the references at the end of this book for further reading, and see the glossary to acquaint yourself with geologic terms used in the book.

Using This Book: Maps, Locations, and Tools

I-5 is the artery providing access to many of the sites in this book. When appropriate, the directions to the sites begin at this highway. However, some places will require hours and miles to reach if you start from I-5.

The Getting There maps at the beginning of each vignette are not intended for detailed navigation. For that, a 7.5-minute (1:24,000 scale) US Geological Survey topographic map or a Green Trails map may be the best source. Excellent aerial photography is now available via the Internet. Programs such as Google Earth and NASA World Wind have proven to be indispensable for teaching geology. The appendix includes a table with the coordinates of important locations in this book. I determined most locations using a handheld global-positioning receiver and obtained others from Google Earth.

Some basic tools will help you more fully enjoy Washington's geology. I recommend the following: a hand lens of at least 10x magnification, a geologist's hammer (not a carpenter's hammer, which may be poorly tempered and can shatter against rock), safety glasses, and a garden trowel or short-handled hoe. Weathered rock surfaces may be coated with a thin veneer of microscopic algae and lichens that obscure the minerals and structures of the rock; hammers and trowels are useful for making fresh rock surfaces to examine by eye or with a hand lens. But do not use hammers or trowels in state parks or national parks and monuments. It is illegal to collect rocks there, too. Always be discrete when you are banging away on an outcrop. Look around for a recently broken rock surface first before making a new one; don't smash outcrops any more than necessary; and wear eye protection. The outcrop you are bashing away at may be of particular interest to geologists in the future or the site of frequent field trips. When in doubt, put your hammer away. If you do break rocks, leave some broken pieces behind for others to examine. It's better to spend your time making observations and thinking about what you see than to wield your hammer aimlessly.

Public lands often require a pass for parking. Passes may be available at or near field trip sites or online. Passes to national parks and monuments can often be purchased at the entrance.

UNITS OF MEASURE AND TIME

This book is for an audience interested in science. If readers read any of the journal articles I cite, they will likely not find English measurements. Consequently, I've included metric units in parentheses after English units.

The age of rock formations or other geologic rock units is often referred to in millions of years, as in this rock formed 5.35 million years ago. The age is gauged from the present. For historical dates I use the nonreligious BCE ("before the common era") and CE ("common era," the one we live in today) rather than the Christian-based BC and AD. For example, an event in 1400 BCE happened 3,415 years before the year in which this book was published, and 1969 CE is the same year as the first moon landing.

ACKNOWLEDGMENTS

A host of geologists reviewed the individual vignettes for accuracy. Thanks to Western Washington University friends and colleagues Scott Linneman, George Mustoe, Scott Babcock, Ned Brown, Sue DeBari, and Don Easterbrook. Sadly, two of them, Drs. Chris Suczek and Maury Schwartz, did not live to see this book in print. Reviewers from the US Geological Survey were Brian Atwater, Wes Hildreth, Kevin Scott, Jim O'Connor, Carl Thornber, Guy Gelfenbaum, Carolyn Driedger, and Russ Evarts. Other reviewers were Stephen Reidel (Pacific Northwest National Laboratory), Stephen Slaughter and Joe Dragovich (Washington State Department of Natural Resources), Doug McKeever (Whatcom Community College), John Clague (Simon Fraser University), Pat Pringle (Centralia Community College), and Jon Riedel and Paul Kennard, geologists with the National Park Service. Connie Soja (Colgate University) and Anton Oleinik (Florida Atlantic University) helped identify fossils in vignette 14. Any inaccuracies are solely my responsibility.

David B. Williams, author of *Stories in Stone: Travels Through Urban Geology*, took me on a personalized guided tour around downtown Seattle and shared his wealth of knowledge about building stones.

Ben Baugh, Joe Goshorn-Maroney, and Graham Andrews, geology students at Western Washington University, made the background topographic images for the Getting There maps. Thanks to Mike Larrabee (North Cascades National Park) for procuring recent data for the Nisqually Glacier recession figure in vignette 10.

John Scurlock, commander of the Mount Baker Volcano Research Center's air wing, provided aerial photos as well as artistic photography and good company inside a totally deserted but thoroughly sodden Ape Cave. Thanks to the many other friends and helpful strangers who contributed photos. Bob and Adena Mooers, Keith Kemplin, Peter Scherrer, Scott Linneman, and Jenevive Delazzari served as field testers. Many other nongeologist friends read chapters and asked hard questions. I am profoundly grateful to James Lainsbury, my editor at Mountain Press Publishing, who cut me a lot of slack as a first-time author.

I extend very special thanks to two individuals: Kevin Scott (US Geological Survey), my colleague during field studies of the recent eruptive history of Mount Baker, for his unstinting encouragement, both to take on this writing project we call "Stompin'" and to persevere for three years to completion. My wife, Kim Brown, made sure there was something to nourish me besides beer, coffee, and chips; I was happy to share the latter with my cat, Chico.

EON	ERA	PERIOD	EPOCH	Estimated age (millions of years before present)	Relative timescale if converted to calendar year	MONTHS
Phanerozoic	Cenozoic	Quaternary	Holocene			Dec.
				0.01	Dec. 31, 11:45 p.m.	
			Pleistocene			
				2.6	Dec. 31, 7:08 p.m.	
		Neogene	Pliocene			Nov.
				5	Dec. 31, 11:50 a.m.	
			Miocene			
				23	Dec. 28	
		Paleogene	Oligocene			Oct.
				34	Dec. 27	
			Eocene			
				56	Dec. 25	
			Paleocene			Sept.
				66	Dec. 24	
	Mesozoic	Cretaceous	Late			
				100	Dec. 20	
			Early			Aug.
				145	Dec. 17	
		Jurassic				
				201	Dec. 10	
		Triassic				July
				252	Dec. 5	
	Paleozoic	Permian				
				299	Dec. 1	
		Pennsylvanian				June
				323	Nov. 26	
		Mississippian				May
				359	Nov. 25	
		Devonian				
				419	Nov. 19	
		Silurian				April
				444	Nov. 17	
		Ordovician				
				485	Nov. 10	
		Cambrian				March
				541	Nov. 3	
Proterozoic						Feb.
				2,500	Apr. 22	
Archean						Jan.
				3,600	Jan. 1	

Morton Gneiss: the oldest known rocks in the United States

The geologic timescale, representing only rocks found in the United States, compared to a calendar year; roughly 80 percent of Earth's rock history is represented in this chart. (You can see Morton Gneiss in building stones in vignette 14.) The combined Neogene and Paleogene periods were formerly known as the Tertiary period; this older term is still commonly used. The geology described in this book is Late Cretaceous or younger; much of it is Quaternary. (Adapted from a drawing in Molenaar 1988.)

THE GEOLOGY OF WESTERN WASHINGTON

From the crest of the Cascade Range westward to the Pacific, and from the Columbia River north to the Canadian border, western Washington hosts a fantastic variety of geology. Here are volcanoes, some very young and others ancient and preserved only as deeply eroded scraps. Rocks from the ocean floor are now exposed on dry land (vignettes 15 and 16). Long-extinct flora and fauna of subtropical floodplains are preserved in sedimentary rocks (vignette 21). Exotic metamorphic rocks, which formed far from Washington, compose the deep roots of the North Cascades (vignette 18). Masses of magma that cooled deep in Earth's crust are now craggy mountain peaks (vignette 19). At least six glacial periods left a veneer of deposits in the northern half of the state (vignettes 11, 15). Geology here is active and very much alive: volcanoes periodically erupt, showering their surroundings with ash (vignettes 2, 7, 8, 22); earthquakes shake Earth's surface and the constructions of humans, sending tsunamis ashore to wreak havoc (vignette 4); alpine glaciers shed their bouldery burdens and sometimes send forth great floods of water (vignette 10); hillsides collapse (vignette 1); and streams erode sediment and rocks and redeposit them downstream (vignettes 3, 10).

Much of this geology is concealed beneath a dense vegetative hell, a gift from the gray skies and rain for which the west side of the state is famous. To piece together the geologic story, rock detectives must first find scattered rock outcrops amidst the greenery and then attempt to relate the rocks with each other. Sometimes, after much effort and study, it is apparent that there isn't a simple correlation. Faults hidden by dense undergrowth juxtapose rocks unrelated in time and space. Profound glacial erosion removed large portions of the geologic record or buried it beneath clay and boulders. Roadcuts often provide the best geologic exposures, and certainly the easiest access to much of the west side's geology.

Newer technology, such as Lidar (vignette 6), is allowing geologists to peer under the vegetative mat for faults and other geomorphic features. Though such groundbreaking technology is accelerating our

TIME	EVENTS	VIGNETTES
20 million years from now	The Juan de Fuca Plate is completely subducted; volcanism in the Cascades ends.	Will books exist? Will humans?
coming soon...	The next big Cascadia earthquake and the next volcanic eruption.	future editions
20th century to present	Global climate change induces the rapid recession of alpine glaciers and outburst floods.	10
May 18, 1980 and after	Eruptions of Mount St. Helens	8
20th century	Building stones are imported from around the world and are used in downtown Seattle. Humans affect sedimentation at the mouth of the Columbia River.	3, 14
January 26, 1700	Magnitude 9 Cascadia earthquake.	4
circa 1480	Mount St. Helens erupts the W tephra.	7
early to mid-1400s	Bonneville landslide.	1
1st century CE (?)	The Cave Basalt erupts at Mount St. Helens.	7
5,600 years ago	The Osceola Mudflow from Mount Rainier reaches the Puget Lowland.	9
post-Fraser glaciation and continuing today	Snoqualmie Falls migration; Dungeness Spit formation and development; honeycomb weathering in the Chuckanut Formation.	13, 17, 20
post-Fraser glaciation	The Mima mounds form.	6
20,000 to 15,000 years ago	The Fraser glaciation occurs in the Puget Lowland. Floods from Glacial Lake Missoula reach the sea.	2, 11, 12
57,000 years ago	The Beacon Rock volcano erupts.	2
late Pleistocene	An unconformity develops at Beach 4.	15
300,000 years ago	The Table Mountain andesite erupts at Mount Baker.	22
496,000 years ago	The Burroughs Mountain andesite erupts at Mount Rainier.	9
Pleistocene	Multiple glaciations occur in the Puget Lowland and North Cascades.	11, 15, 22
1.15 million years ago	The Kulshan caldera erupts and collapses in the Mount Baker Volcanic Field.	22
15.5 million years ago	The Columbia River Basalt flows down the ancestral Chehalis River.	5
22 to 16 million years ago	Turbidites are deposited off the coast; Olympic Subduction Complex rocks are uplifted and exposed along the coast.	15
45 million years ago	Earliest plutons intrude the North Cascades; the Skagit Gneiss forms through metamorphism.	18, 19
52 million years ago	*Diatryma* walks western Washington's floodplains.	21
Eocene epoch to present	The Olympic Mountains are uplifted.	16
58 to 31 million years ago	The Crescent Formation basalt erupts.	16
100 million years ago	Wrangellia is accreted to the western edge of North America.	18
230 to 185 million years ago	North America begins its westward motion, and terranes begin to be accreted to its western margin.	introduction
3.524 billion years ago	The oldest components of the Morton Gneiss are erupted.	14
4.54 billion years ago	Earth coalesces from a dust cloud.	

The timeline of major geologic events in western Washington, and the vignettes that explore them. The earliest events are at the bottom.

understanding of Washington, the information it provides is only a guide; geologists must still visit places of interest to understand their geology.

Assembling this diverse geology into a reasonably coherent and continuous whole is a fine challenge, and plenty of work remains. Geology is a good field for young natural historians who like puzzles. Geologists will continue to fine-tune our understanding of the geologic history and structure of western Washington for generations to come.

Plate Tectonics:
The Cascadia Subduction Zone

The theory of plate tectonics (from the Greek *tekton*, meaning "builder") helps unify the geologic story of western Washington and the entire planet. According to the theory, the outer layer of the planet consists of rigid plates of brittle rock that move about Earth's surface. The plates are part of a composite layer called the lithosphere, which consists of the thin crust (igneous, sedimentary, and metamorphic rocks composing Earth's outermost layer) and the solid uppermost portion of the underlying mantle, which is mostly peridotite, a rock composed largely of the mineral olivine. The Earth's fifteen principal plates slide around on the hot, weak, ductile rock of what's called the asthenosphere, also a part of Earth's mantle. Another thirty-five or so microplates, some quite tiny, have been recognized. The crust ranges up to 60 miles (100 km) thick. The rigid mantle component of the lithosphere is 30 to 70 miles (50 to 110 km) thick, whereas the plastic asthenosphere is about 120 miles (200 km) thick.

The process that drives plate motions is a subject of some contention. Many geological theorists hold that enormous lateral currents, or convection cells, carry the plates along piggyback-style in the mantle. Convection cells work similarly to the circulation of hot fudge in a cook pot: hot rock rises upward because it is less dense, hence more buoyant, than the adjacent mantle, and cool rock sinks. Earth's core heats the mantle, causing great upwellings, or plumes, of molten — or at least very plastic — rock to rise. When the rising mantle reaches the brittle lithosphere, it cools and circulates laterally until it has cooled enough to be denser than the mantle below. The rock then sinks back into the mantle, perhaps as far as the core-mantle boundary, completing the convection cell. Heating by Earth's core perpetuates the circulation.

Others maintain that the core doesn't give off sufficient heat to drive this motion. One alternative theory posits that oceanic plates slide downhill from their highest elevations, at spreading centers, toward their lowest points, at subduction zones. The descent of the cold, dense plates at subduction zones pulls the rest of the plate along in a mechanism called slab pull; thus, the theory argues, it is the motion of the plates that drives passive circulation in the mantle. Regardless of the mechanism of movement, all geoscientists agree that the lithosphere moves laterally on the asthenosphere, driving the geology we see at the surface.

There are two types of plates: oceanic and continental. Oceanic plates are thin, rarely more than 6 miles (10 km) thick, and made principally of the dense volcanic rock basalt. Oceanic plates form topographic lows on Earth's surface that are flooded by the oceans. Continental plates are up to 60 miles (100 km) thick and mostly made of lower-density rocks, such as granite. Both types of plates are less dense than the mantle, so they float above it, just as oil floats on water. Continental plates float higher than oceanic plates and are mostly above sea level because they are less dense. Washington and the rest of the continent are on the North American Plate. Just offshore to the west lies the oceanic Juan de Fuca Plate, a microplate; it is a leading actor in western Washington's geo-drama.

There are three types of boundaries where plates meet: divergent, convergent, and transform. Because plate boundaries are places where rock bodies move relative to each other, these boundaries are also faults. At divergent boundaries, also known as spreading centers, plates pull apart. Most of the planet's spreading centers are on ocean floors. Convergent boundaries occur where plates collide, head-on or at an oblique angle. In the typical case, an oceanic plate slides beneath a continental plate or another oceanic plate at a thrust fault; this process is called subduction. The Pacific Ring of Fire is a nearly continuous convergent boundary where oceanic plates subduct beneath continental plates or other oceanic plates.

In places where two continental plates collide, subduction doesn't work due to the thickness and similar densities of the colliding plates. Instead, high mountain ranges are pushed up. The three great mountain ranges of central Asia (Himalaya, Karakoram, and Hindu Kush) are the classic example of this type of boundary. The mountains are forming as the Indian Plate slowly pushes northward and collides with the Eurasian Plate, crumpling the crust in a titanic slow-motion process. What is slow motion? At the onset of the plate collision 45 million years ago, the relative northward motion was 1.6 inches (4 cm) per year. Today the relative rate is much slower, 0.15 inch (0.4 cm) per year.

Transform boundaries are found where plates slide laterally past each other. The San Andreas Fault forms the boundary between the North American and Pacific Plates in California and is a familiar example.

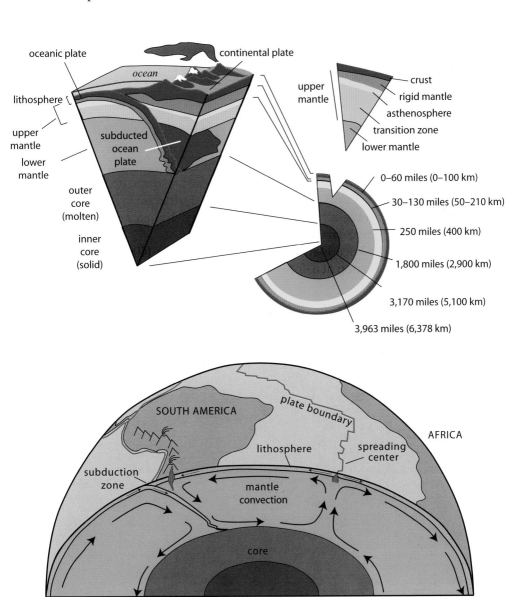

An interpretation of Earth's structure and the convection-cell theory of plate tectonics. Hot rock in the mantle rises toward mid-ocean spreading centers and moves laterally at the boundary with the lithosphere; tectonic plates are carried along by the hot rock. Mantle rock cooled by proximity to the planet's surface sinks back into the deep mantle at subduction zones, carrying oceanic plates with it. Magma reaches the surface at mid-ocean spreading centers and cools, adding new material to the crust.

Western Washington is part of a seismically active geologic province known as Cascadia. It includes the southwest corner of British Columbia, western Oregon, and the northwest corner of California. Cascadia is defined by the Cascadia Subduction Zone: from northern California to southern British Columbia, the eastward-moving Juan de Fuca Plate meets the leading edge of the North American Plate along this convergent boundary, which is about 80 miles (130 km) off the coast of Washington. Because it consists of denser rock, the Juan de Fuca Plate is subducted under the continental plate.

The 600-mile-long (1,000 km) Cascadia Fault is a planar structure running the length of the boundary between the two converging plates, from Earth's surface to the base of the lithosphere. The plates in the subduction zone don't slide smoothly past each other. Motion is herky-jerky. Stress builds up in a "locked zone" between the plates due to friction and increases until the fault fractures, allowing the Juan de Fuca Plate to slide a little farther — maybe 10 feet (3 m) — beneath the continent. This seemingly slight amount of slip can generate huge earthquakes (vignette 4).

The Juan de Fuca Plate begins its life 250 miles (400 km) off the coast of Washington at a spreading center, or rift, called the Juan de

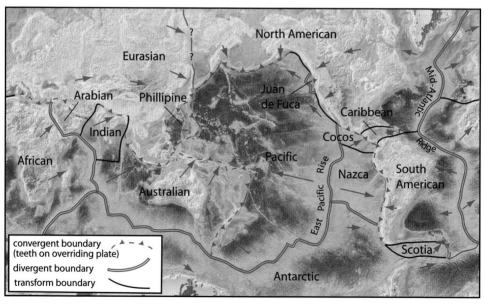

The principal plate boundaries. Red arrows indicate relative plate motion, and the length of the arrows reflects the relative velocity; long arrows indicate faster motion. The Nazca and Pacific Plates move the fastest, about 4 inches (10 cm) per year. Light-green areas are shallow submarine shelves at continental margins. The "teeth" on convergent boundaries lie on the overriding plate.

Fuca Ridge. The 600-mile-long (1,000 km) ridge is oriented slightly east of north-south. The small Juan de Fuca Plate and the huge Pacific Plate diverge here about 1.6 inches (4 cm) per year. There is little to no crust covering the mantle at this rift, which means the mantle rock below is under less pressure. This decreased pressure allows about 15 percent of the solid but very hot mantle rock to melt and form magma (the melting point of solid rock is reduced if pressure decreases). The magma, being principally a fluid, has lower density than the surrounding solid mantle rock, so it rises into the lower-pressure spreading center, bulging the spreading center upward to form a chain of volcanic vents.

The thinnest part of the oceanic crust underlies a 5-mile-wide (8 km) valley running along the crest of much of the Juan de Fuca Ridge; the valley marks the divergent plate boundary. Here is where magma leaks out of the mantle to periodically erupt as lava, covering the valley floor with young lava flows that become new oceanic crust. This process occurs at other divergent plate boundaries; the entire rocky crust beneath the world's oceans consists of basalt lava erupted at mid-ocean spreading centers. Most of the volcanism along the Juan de Fuca Ridge occurs periodically at scattered sites in the valley, but some of it is concentrated in large volcanoes. Lava flows have been caught on underwater cameras and instruments at Axial Volcano, the most active of these centers on the Juan de Fuca Ridge. This underwater mountain rises 3,600 feet (1,100 m) above the seafloor west of the central Oregon Coast. Its summit is 4,600 feet (1,400 m) below sea level.

The Juan de Fuca Plate, its smaller northern and southern neighbors (the Explorer and Gorda Plates, respectively), and the Cocos Plate (west of Central America) are remnants of the much larger Farallon Plate, which formed the eastern part of the Pacific Ocean basin during the Mesozoic and Cenozoic eras. Nearly the entire Farallon Plate has been subducted beneath North America. A nearly vertical, 250-to-370-mile-wide (400 to 600 km), crumpled remnant of this plate extends 500 to 1,200 miles (800 to 2,000 km) down into the hot mantle, sinking at around 0.4 inch (10 mm) per year.

Meanwhile, the North American Plate, with western Washington at its leading edge, migrates west-southwestward about 1 inch (2.5 cm) per year. This enormous plate extends eastward to the center of the Atlantic Ocean, where another divergent plate boundary, the Mid-Atlantic Ridge, runs the length of that great ocean. The ridge divides the North and South American Plates from the Eurasian and African Plates. At first glance, the migration rate of the North American Plate appears insignificant. However, North America has

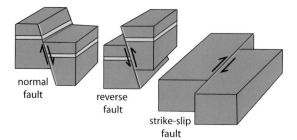

The three styles of faults. Divergent plate boundaries are characterized by normal faulting; subduction occurs along low-angle reverse faults called thrust faults; transform boundaries are along strike-slip faults.

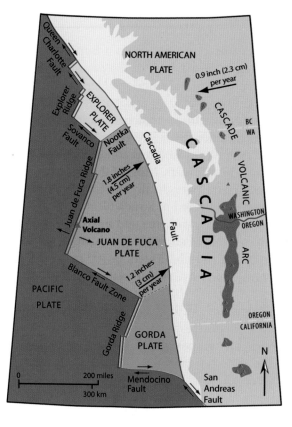

Cascadia's tectonic setting. The Cascadia Fault marks the boundary between the North American Plate and the Juan de Fuca Plate and its smaller neighbors. The combined rate of relative movement along this fault at the surface is 2.5 inches (6.3 cm) per year. The Cascade volcanoes parallel the subduction zone.

been migrating westward at about that rate for the last 200 million years. During that time the plate has moved generally westward thousands of miles, and the Atlantic Ocean has grown wider.

Chains of volcanoes, or volcanic arcs, grow on the crust above subduction zones (vignettes 2, 7, 8, 22). The Cascade Volcanic Arc, comprising the volcanoes of the Cascade Range, is a good example. The Juan de Fuca Plate, which begins its descent under North

America at the Cascadia Fault, is 50 to 60 miles (80 to 100 km) below the surface along the line defined by the volcanoes. At that depth, seawater saturating the descending slab of basalt boils out of the rock and rises into the hot but solid crystalline mantle between the slab and the bottom of the continental plate. Water lowers the melting point of rocks by weakening bonds between molecules, so a small fraction of the mantle melts. The melted rock, or magma, filters upward between crystals in the mantle rock and accumulates near the base of the continental plate. As magma slowly cools, minerals crystallize and segregate due to their contrasting densities. Heavy atoms, such as iron and magnesium, concentrate in dense minerals, such as olivine and pyroxene, which sink toward the bottom of the magma. The remaining magma becomes enriched in lower-density elements, such as oxygen, silicon, and aluminum, and rises upward into the solid continental crust. Some of this magma will accumulate and form plutons (vignette 19); some of it will continue upward and erupt from the Cascade volcanoes.

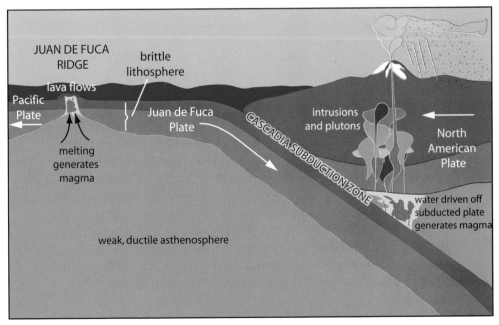

The Juan de Fuca Plate moves eastward from the Juan de Fuca Ridge. The water-saturated oceanic plate subducts beneath the westward-moving North American Plate. Hot water is driven upward into the mantle, causing some of it to melt. The resulting magma rises into continental crust and either accumulates beneath the surface as plutons or erupts from the Cascade volcanoes.

E Pluribus Unum: Assembling Washington

All of the lithosphere beneath Washington (and the rest of the continent west of the Rockies) originally formed elsewhere, only becoming part of the North American Plate in the latter part of the Mesozoic era. Dozens of terranes, blocks or fragments of crust bounded by faults and composed of rocks sharing a similar history, make up the western portion of Washington. Some writers refer to these bits of crust as "suspect" or "exotic" terranes because they wandered long distances across the planet's surface.

The supercontinent Pangaea was an amalgamation of most of Earth's landmass during late Paleozoic and mid-Mesozoic time. Around 230 million years ago Pangaea began to break apart into subordinate tectonic plates, including most of what has become the North American Plate. The main north-south rift that Pangaea separated along widened about 1 inch (2.5 cm) per year and was eventually flooded by the Panthalassa Ocean, which had surrounded the supercontinent. Rifting is happening today between the African and Arabian Plates, with the Red Sea widening between them.

The developing North American and Eurasian Plates were fully separated by a narrow Atlantic Ocean sometime between 60 and 55 million years ago. At the time, the western margin of North America was in the vicinity of today's Washington-Idaho boundary. As the Atlantic opened, the North American Plate moved westward. The western margin (the leading edge) of the plate overran and subducted the thin oceanic plates beneath the narrowing Panthalassa Ocean; today we call the remnant of this ocean the Pacific. Scattered about the ocean were volcanic island chains, similar to the Solomon or Mariana Islands, and small plates that included larger landmasses, similar to Japan or New Guinea. Plate movements carried these small plates here and there around the Pacific before they were finally delivered, conveyor-belt-style, to the subduction zone at North America's western doorstep. These islands and small continents were too thick and buoyant to be subducted (vignette 18), so they were accreted to the continental margin. The North American Plate grew incrementally by adding these far-traveled terranes. Accretion took place off the coast, at the subduction zone's outer margin, while the intervening seafloor sediment was crushed in between.

Popular interpretations of tectonics often use terms such as "slam" and "crash" to describe plate collisions. It is a mistake to envision a fleet of islands grandly sailing into view, appearing ever closer until they dock against the continent. Accretion events happened

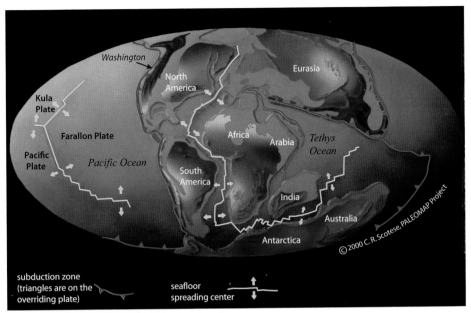

Earth's tectonic setting 94 million years ago. Pangaea has broken into separate plates along the spreading center that will form the Atlantic Ocean. The Panthalassa Ocean has divided into the Tethys and Pacific Oceans. Continued subduction off the west coast of North America will bring the spreading centers in the Pacific closer to the continent. (Adapted from a drawing by C. R. Scotese, ©PALEOMAP Project.)

too slowly to be noticed by the dinosaurs and their successors who roamed the continent. Other than the occasional large earthquake, accretion goes largely unnoticed. Furthermore, the initial contact occurred at the subduction zone, well off the coast. Accreting lands were not even visible from shore until considerable compression of the intervening seafloor occurred.

The first episode of accretion in what would become Washington occurred 185 to 170 million years ago, during the Jurassic period. A small plate some 600 miles (1,000 km) long and consisting of volcanic islands and larger landmasses lay in the ocean west of North America. Subduction swallowed the thin oceanic lithosphere separating the plate and the coast of North America until they were drawn into collision. In the final stages of collision, the intervening limestone, shale, and sandstone on the seafloor were shoved eastward and folded and raised above sea level to form the tortured sedimentary rocks of the Columbia Mountains (part of the Rocky Mountains) in the northeast corner of Washington. Over millions of years the small plate became attached to the continental plate

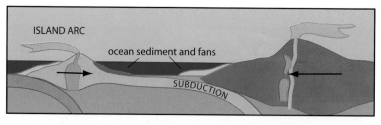

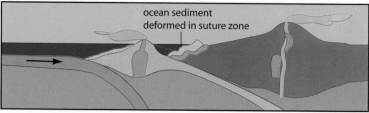

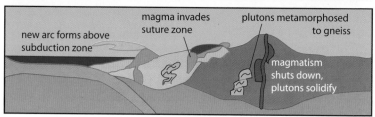

Terrane accretion. Top: Tectonics brings an island arc toward a larger plate. Center: Oceanic sediment is deformed as the ocean basin narrows. Bottom: The island arc has accreted, and terranes are stacked atop each other along thrust faults. The pressure and heat of the collision metamorphoses rocks within the accreted terrane; magma rises into the suture zone.

and is now known as the Intermontane Superterrane, which consisted of a number of smaller terranes that had previously been accreted together. Rocks associated with the superterrane extend from the Yukon south along the west edge of the Canadian Rockies to the Okanogan Highlands in northeast Washington. The superterrane may extend farther south, buried by the thick Columbia River Basalt Group lava flows.

The North American Plate continued to move westward, and rocks that are today found in western Washington entered the picture. In Late Jurassic to Early Cretaceous time, new terranes, primarily seafloor rocks, sidled up to North America, piggybacking on the subducting Farallon Plate. These terranes belong to the Insular Superterrane, which was accreted to the western margin of the Intermontane Superterrane. The Insular Superterrane extends from central Alaska to southern Vancouver Island. Some of its terranes may have been assembled tens or hundreds of millions of years before they ever came close to North America, perhaps prior to the assemblage of Pangaea.

Wrangellia, the largest of these terranes, is a thick accumulation of basalt and marine sediments that formed an extensive plateau on the seafloor. Fossils in the sedimentary rocks under, between, and on top of the lava flows, and the alignment of magnetic minerals in the basalt, suggest that the basalt erupted 231 to 225 million years ago (Late Triassic time) at equatorial latitudes far west of North America. Wrangellia is so vast that it includes all of Vancouver Island, the Queen Charlotte Islands, and a large swath of interior Alaska, as well as much of the seafloor off western Canada.

Around 160 million years ago, during the Late Jurassic, Wrangellia had moved to the margin of North America north of present-day Washington. Rocks of the future Methow Terrane (vignette 18), today exposed on the east side of the North Cascades, were accumulating as sediment just offshore.

By 130 million years ago, subduction of the Farallon Plate at an oblique angle dragged Wrangellia to a position off the coast of southern British Columbia. Subduction generated magma that penetrated the developing suture between Wrangellia and the Intermontane Superterrane. The magma would become the older plutons of the Coast Plutonic Complex, the huge mass of intrusive and metamorphic rocks found in the northernmost Cascades, the Coast Mountains of British Columbia, and the Kluane Range in southwestern Yukon. The complex is widely considered the world's largest belt of plutonic rocks. In the meantime, terranes that would eventually be incorporated into the Northwest Cascades System (see vignette 18) were accreting to the continental margin to the south, perhaps in California, or even farther away at the latitude of Baja California (the controversial Baja British Columbia hypothesis is discussed below).

By about 100 million years ago, Wrangellia had completed its journey and was a part of the continent. The narrow sea floored by the Methow Terrane was closed as subduction-driven compression squished and uplifted the marine sediment. Wrangellia protruded westward from the continent; the protrusion is preserved today in the westward jut of Vancouver Island, the southern edge of the terrane. To the south, the future Northwest Cascades System terranes began to slide northward up the coast. These terranes collided with the south end of Wrangellia and, according to some, were stacked up by thrust faulting during the collision to form the rocks in the San Juan Islands and the northwestern Cascades. Subduction continued, and great volumes of magma added thickness to the Coast

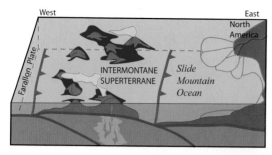

185 million years ago: The accretion of the Intermontane Superterrane to North America drives Slide Mountain Ocean sediments onto the continent and raises the Columbia Mountains in northeast Washington.

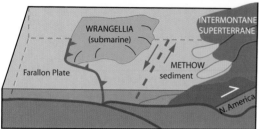

160 million years ago: The Intermontane Superterrane has been thrust over the margin of North America. Offshore, Wrangellia (submarine) has been accreted and has begun to slide south. The future Methow Terrane is deposited as sediment on the ocean floor.

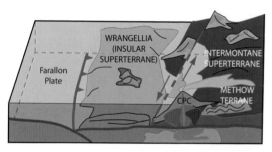

130 million years ago: Wrangellia is offset farther south. The closing of the Slide Mountain Ocean basin accretes and crumples the Methow Terrane. The Coast Plutonic Complex (CPC) develops in the suture between Wrangellia and the Intermontane Superterrane.

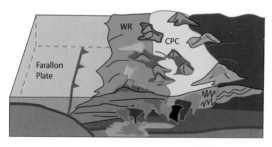

100 million years ago: Wrangellia (WR) is fully accreted. Older plutons in the Coast Plutonic Complex (CPC) are metamorphosed to gneiss.

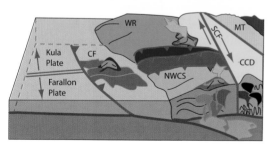

55 million years ago: The Northwest Cascades System (NWCS) terranes have migrated north and been thrust over Wrangellia. Movement along the Straight Creek Fault (SCF) offsets the Crystalline Core Domain (CCD) 100 miles (160 km) south into Washington from its previous home in southern British Columbia; the core had been part of the Coast Plutonic Complex, but the movement along the fault brought it into contact with the Methow Terrane (MT). Crescent Formation basalt (CF) erupts onto the continental shelf as the Kula-Farallon spreading center subducts. This panel is about 100 miles south of the others.

Western Washington accretion sequence. The Baja BC hypothesis is not accounted for in this diagram.

Plutonic Complex and contributed to the metamorphism of the region's bedrock (vignette 18).

Here is where the Baja British Columbia hypothesis comes into the story. Some geologists propose that most of the terranes that form western Washington and British Columbia, and much of Alaska, were accreted to the continent one after the other—and stacked along thrust faults—as far south as today's Baja California, far from their current positions in western Washington. All the accreted terranes of the Insular Superterrane, the intruded material of the Coast Plutonic Complex, and even portions of the Intermontane Superterrane then slid en masse 1,200 miles (2,000 km) up the North American margin along as-yet-unknown San Andreas–style strike-slip faults. The hypothesized and no-longer-active faults could easily be buried beneath younger rocks or could have been destroyed by younger plutons of the Cascade Volcanic Arc.

The hypothesis has remained controversial, hinging on the orientation of magnetic minerals in 90-million-year-old plutonic rocks. When magnetic minerals crystallize in magma, they preserve the direction to the magnetic poles at the time, thus leaving geologists evidence of when and how far north or south of the poles the rocks formed. Some plutons indicate a southern origin, but there is field and paleomagnetic evidence that supports both northern and southern origins for the terranes. Some studies support long-distance transport, others more moderate movement of perhaps 500 miles (800 km). Other studies suggest little to no northward movement at all.

Whether or not the Baja British Columbia idea holds true, all the terranes of the Pacific Northwest were assembled and in place 55 million years ago. The continental margin was west of the Cascades and Vancouver Island, but not as far west as at present; rocks of the Olympic Peninsula were still submarine. Subduction of the spreading center that separated the Farallon and Kula Plates produced the huge outpouring of Crescent Formation basalt (vignette 16) around 53 million years ago. And volcanic activity began in the Cascades about 35 million years ago.

All the older rocks are, or were, covered by various younger geologic units. These include the Chuckanut Formation (vignettes 20 and 21) and younger Neogene- and Paleogene-age sedimentary rocks in southwestern Washington and the Puget Lowland, Cenozoic-age volcanics in the southern Cascades, Miocene-age Columbia River Basalt Group flows (vignette 5), and Pleistocene-age glacial sediment (vignettes 3, 11, 17) and volcanics (vignettes 2, 9, 22). Intense erosion during repeated glacial advances stripped

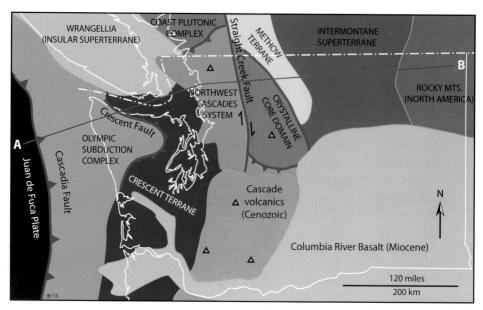

Simplified terrane map of Washington and southern British Columbia. The Crystalline Core Domain (vignette 18) is part of the Coast Plutonic Complex that was moved south on the Straight Creek Fault. Gray colors mark younger units that obscure terrane boundaries. The red line indicates the cross section in the next figure. (Adapted from Tabor and Haugerud 1999 and Wilson and Clowes 2009.)

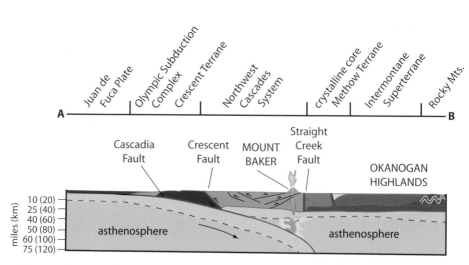

Cross section across the northern edge of Washington from the Juan de Fuca Ridge to the Idaho border. The southern edge of the Insular Superterrane, including Wrangellia on Vancouver Island, is just north of the line of the cross section. The red area represents the preaccretion leading edge of the North American Plate. Cenozoic surface deposits are not shown. (Adapted from Brown and Dragovich 2003 and Wilson and Clowes 2009.)

away these young rocks in the northern tier of the state to reveal the ancient rocks below, allowing geologists to study them and piece together this remarkable geologic history. Beyond the reach of glaciation in southern Washington, the older volcanics of the Cascades still obscure Mesozoic-era geology.

Today, rocks of the Olympic Peninsula are well above sea level and rising (vignettes 15 and 16), lifted because subduction continues along the coast. Subduction leads to continued Cascades eruptions (vignettes 7, 8, 9) and huge earthquakes (vignette 4). Agents of erosion, subtle (vignette 3) and not so subtle (vignette 1), continually change the landscape.

Geology Marches On: The Future

Sedimentation by streams and glaciers continues in western Washington, as do volcanic eruptions. New rocks may someday cover today's surface rocks, and unrelenting erosion will remove some of the surface rocks to expose deeper ones. Rivers will change channels, and waterfalls will be worn away (vignette 13). Periodic glacial advances out of Canada's interior are likely. Earthquakes and faults will continue to break rocks and locally raise or lower Cascadia's land surface on a small scale. Uplift will raise the Cascades and erosion will counteract it.

Predicting the bigger picture — the rearrangement of tectonic plates — is riskier business. Let's assume that today's tectonic regime continues. Subduction under western Washington will continue until the Juan de Fuca Plate disappears beneath North America in the swan song of the Farallon Plate and its descendants. From its eastern margin at the Cascadia Subduction Zone to the Juan de Fuca Ridge, the oceanic plate is only 190 miles (300 km) wide. The continental plate is moving west-southwest at around 1 inch (2.3 cm) per year. At that rate, subduction will continue for another 12 million years (do the math yourself). There are no offshore landmasses such as island chains, large seamounts, or submerged plateaus between the ridge and the North American Plate, so we can't expect any new terranes to come shuffling along. Once the last fragment of the Juan de Fuca Plate subducts, the plate tectonic regime will change to a San Andreas–style strike-slip fault. That fault system is migrating up the coast from northern California.

What about future volcanism? The Juan de Fuca Ridge is oriented obliquely to oncoming North America. As the spreading center at the ridge subducts, the descending Juan de Fuca and Pacific Plates will continue to diverge, possibly creating what's called a slab window (see vignette 16). If this occurs, huge volumes of magma could erupt and a great new volcanic province — along the lines of the Olympic Peninsula's Crescent Formation basalt — might develop on the continental margin.

Finally, what will happen to the Cascade Range? As the last of the Juan de Fuca Plate sinks below the continent, magma will cease to intrude the crust and Cascade volcanism will subside. The mountains will stop rising since they no longer will be fed by additional plutonic rocks. Erosion will outpace mountain building, and the rugged Cascade Range will slowly wear down to hills and, finally, a plain. Western Washington will once again be a region of low relief extending far to the east, as it was in Eocene time (vignette 21). The North American Plate will slide farther to the southwest, perhaps even outside the temperate zone. Then, even during cold periods, glaciers will not form on the stubs of the Cascades.

On a more global scale, the Pacific Ocean is getting smaller as you read this, whereas the Atlantic Ocean is widening as the North and South American Plates migrate west-southwesterly, away from the Mid-Atlantic Ridge. Eurasia sweeps eastward. The western sliver of California, part of the northwest-moving Pacific Plate, drives northward along the San Andreas Fault toward Alaska; in 10 million years Los Angeles will be at the same latitude as San Francisco, and 50 million years later it will have migrated past the western edge of Cascadia to the south edge of Alaska at the Aleutian Islands, where it will, finally, be accreted to North America.

The eastern margin of Eurasia will eventually swallow the oceanic Philippine Plate and accrete the islands it contains before colliding with the north-moving Australian Plate. The hodgepodge of microplates (for example, the North and South Bismarck Plates, the Woodlark Plate, and the Caroline Plate) east of the Philippines Plate and north of the Australian Plate will meet the same fate. In the ultimate act of accretion, the mighty Eurasian and American plates could converge at considerably less than a snail's pace and form a new supercontinent. What is now western Washington would be crushed in the giant vise they create. A huge mountain range could rise along the suture of these plates just as the Himalaya Mountains are rising today from the collision of India and Eurasia. If they have survived erosion, the rocks we can visit today will be buried, uplifted, folded, faulted, and metamorphosed until they are virtually unrecognizable.

Other tectonic futurists predict that about 50 million years from now, the westward motion of the North American Plate will reverse. Subduction in Cascadia will end, along with its volcanism.

The western margin of North America will become the trailing edge of the plate, as the Atlantic coast is today. A new subduction zone will open off the east coast of the two American plates. The Pacific will widen. Eurasia, Africa, and the Americas will again reunite into something like Pangaea, swallowing the Atlantic in the process. In short, the global cycle of supercontinent formation followed by supercontinent breakup will continue. Anyone want to lay long odds?

Petrology and Looking at Rocks

The Cascade volcanoes are Cascadia's geologic icons. A disproportionate number of this book's vignettes examine volcanic rocks or their subsurface plutonic equivalents. The geologic names for the various igneous rocks visited in these vignettes, such as andesite or granodiorite, appear often. Scientists need to describe, classify, and pigeonhole the things we study. Various methods have been used to classify igneous rocks. Even though in Geology 101 you may learn that dark volcanic rocks are basalt, medium-gray ones are andesite and dacite, and light ones are rhyolite, color is not a good criterion for differentiating rocks. Colors vary widely based on degree of weathering and a host of other factors.

Igneous petrologists, the geologists who study rocks derived from magma, define rocks based on chemical composition. Sure, igneous rocks can be preliminarily grouped visually based on the combination of minerals they contain. This method is not, however, an infallible way to categorize and understand their magmatic origins. Are there green olivine crystals in that hunk of lava in your hand? Then maybe the rock is basalt, but it could be andesite. Does it contain black needles of hornblende? Well then, it is almost certainly not basalt, but it could be andesite, though more likely it's dacite or rhyolite. But . . . which?

Petrologists use mineral composition as a rough guide for classification until they perform refined lab analyses to determine how much silica a rock sample contains. Silica, the most common chemical constituent of igneous rocks, consists of silicon and oxygen atoms. The proportion of silica an igneous rock contains determines how it is grouped. Volcanic rocks with less than 52 percent silica by weight are called basalt. Andesite contains 53 to 63 percent silica, dacite 63 to 72 percent, and rhyolite more than 72 percent. Petrologists may further subdivide these major categories, so a lava sample at the low end of the andesite scale, with 52 to 57 percent silica, is called basaltic andesite.

oxidized surface

fresh surface

vesicular and glassy scoria

hydrothermally altered; rich in yellow sulfur crystals

fresh surface

discolored weathering rinds

The mineral composition of these andesite samples from Mount Baker is similar. They illustrate why color is not a reliable guide for rock identification. Sulfur-rich gases in Sherman Crater hydrothermally altered the white one; it was once a perfectly respectable dense, gray andesite. Elephant for scale (scale is important in geologic illustrations).

	less than 52% silica	52%–57% silica	53%–63% silica	63%–72% silica	68%–72% silica	more than 72% silica
VOLCANIC ROCKS (erupted)	basalt	basaltic andesite	andesite	dacite	rhyodacite	rhyolite
PLUTONIC ROCKS (cooled in crust)	gabbro	diorite		granodiorite		granite

A generalized classification of igneous rocks.

With your back to the light, hold a hand lens and sample close to your eye to get the largest field of view.

Igneous rocks provide good opportunities to learn to recognize some of the silicate minerals—those composed of silica bonded to other elements. You'll need a hand lens, a freshly broken rock surface, and a good mineral identification guide. A few are listed in the "General Reading and Geologic Maps" section of the references. Igneous rocks are divided into two classes: plutonic, those that form beneath the surface, and volcanic, those that form at the surface. The former, like granodiorite or granite, are composed of interlocking mineral crystals that can be seen by the naked eye, whereas volcanic rocks have quite small crystals, sometimes invisible without a microscope or hand lens, in a matrix of microscopic crystals or rapidly quenched magma called volcanic glass.

The size of mineral crystals is determined by the length of time a particular magmatic rock cooled. Plutonic rocks cool for longer periods because they are better insulated within the crust and are often part of a considerable volume of magma. Longer cooling permits the atoms and molecules that make up the various minerals to migrate through the magma and bond together to form chemically stable crystals. Volcanic rocks have small crystals that rarely exceed 0.2 inch (5 mm) across because the magma that produced them cooled quickly on the surface. Metamorphic rocks with a coarse crystalline structure, such as Skagit Gneiss (vignette 18), are good for learning minerals, too. The tremendous pressures that metamorphosed these rocks made the original crystals even larger and easier to identify.

Igneous rocks that formed in subduction zones consist of only a few minerals. Felsic minerals are rich in silica and are generally clear, white, or other pale colors. The principal felsic minerals in western Washington's rocks are two feldspars (plagioclase and orthoclase) and quartz. Calcium- or sodium-enriched plagioclase takes the form of prismatic clear or white crystals; sometimes it is iridescent. Blocky potassium-rich orthoclase is usually white or pink and more opaque than plagioclase. It takes some practice to tell them apart: plagioclase has closely set parallel lines on crystal faces, whereas orthoclase does not. Quartz breaks irregularly, like glass; is colorless, white, or smoky gray; and is more common in plutonic rocks. Unless it grows in a cavity within a rock, quartz does not form the symmetrical, hexagonal, pointed crystal shape we are most familiar with; instead it forms a rather amorphous crystal that grows between other minerals.

Mafic minerals contain elevated amounts of iron or magnesium and are dark colored or black. Common mafic minerals include two pyroxenes, both forming chunky, more or less cubic crystals. Augite is dark green to black, and hypersthene is beer-bottle brown. Hornblende usually grows as slender black prisms; if large enough, you may see that, in cross section, a hornblende crystal is diamond shaped with angles of 60° and 120°. Biotite is a flaky, bronze mafic mineral in the mica family. It can be seen with the naked eye in some plutonic rocks, especially the Golden Horn granite (vignette 19).

How to Date a Rock

Rocks and geologic events can be dated by a variety of methods. Relative ages are based on the relative position of rock layers or units. A sedimentary rock layer that lies atop another is usually the younger of the two. A dike is younger than the host rock it intruded. A rock that includes fragments, or clasts, of other rocks is younger than the rock unit or units that produced the clasts. A metamorphic rock is younger than its parent rock. Relative ages do not tell us the numeric age of a rock. To do that, geologists rely on other methods.

Dendrochronology (tree-ring dating) can be used to date young lahar or flood deposits, moraines, landslides (vignette 1), or earthquake deposits (vignette 4). A small core can be drilled out of a tree grow-ing on the deposit to determine the tree's age in calendar years. Of course, a geochronologist needs to have good reason to believe the tree began its life very shortly after the event that deposited the sur-face being dated. If the outermost rings survive on trees killed by a geologic event, such as a forest submerged by a landslide-dammed lake or earthquake-induced subsidence, the tree rings can be com-pared with those of living survivors to obtain the victims' last year of growth.

Isotopic dating methods are used to date rocks, minerals, and tiny organic fragments mixed with young sediment. An element is identified by the number of protons contained in its nucleus. The number of neutrons in a given element, however, may vary, and this variation defines the isotopes of elements. For instance, a carbon atom always has 6 protons, but it may have 6, 7, or 8 neutrons. The isotopes carbon-12 (6 protons and 6 neutrons), carbon-13 (6 protons and 7 neutrons), and carbon-14 (6 protons and 8 neutrons) refer to this ratio. Carbon-14 is a radioactive isotope, or radioisotope.

Radioisotopes lose subatomic particles from the nucleus until a stable decay product is reached. In the process, the radioisotope becomes less abundant. For example, the stable isotope carbon-12 remains unchanged in an organic material such as a twig, but carbon-14 decays to become the stable element nitrogen-14. Uranium isotopes (uranium-238 and uranium-235) decay to the isotopes lead-206 and lead-207, respectively, at measured rates. The decay rate of radioisotopes is nearly constant, and we know the naturally occurring ratio of isotopes of the same elements; carbon-14 is about 1 trillionth as abundant as carbon-12. By determining the ratio of a radioisotope and a stable isotope in a substance, scientists can calcu-late the length of time that has passed since the material formed. For example, since carbon-14 leaves organic material at a steady rate, by determining the ratio of carbon-14 to carbon-12 in a sample scien-tists can determine its age.

Carbon-14 can be used to date organic matter, such as charcoal and wood fragments, even those smaller than the head of a match, but it is only useful to measure material less than 50,000 years old. After that amount of time, the amount of carbon-14 has decreased beyond the limits of detection by current methods. (See vignette 4 for more on this method.)

The exposed surfaces of rocks can be dated using the ratio of chlorine-36 to its parent elements (vignette 12), but the more typi-cal way to determine the age of rock is to use radioisotopes con-tained within minerals. Uranium-lead isotope dating is one method.

A newer technique that can give very precise results is the argon-argon method, which measures the ratio of argon-40 to argon-39, particularly in the minerals hornblende and biotite. These methods only work for some rocks, and only some of the time: the right minerals with a sufficient amount of the target isotope must be present; the rock can't be so old that the radioisotope is too depleted through decay; and the rock can't have been reheated by a nearby intrusion, because the minerals will have been reheated and may have absorbed additional amounts of the targeted isotope.

If it's so easy, you might wonder, why haven't the ages of all rocks been determined? Let's say you want to date a particular rock, and it contains zircon. Zircon crystals, or zircons, though usually 0.04 inch (1 mm) or less in size, are frequently found in granitic rocks, are resistant to the temperature resetting of their atomic clocks by heat, and can be separated in the lab. Fortunately for geochronologists, small amounts of uranium often occur in the structure of zircon crystals, so the decay of uranium to lead can be measured. To do that, one must first collect enough rock to have a fair chance of obtaining enough zircons to study; that can mean 10 to 100 pounds of uncrushed rock. That's unfortunate if the rock you want to date is deep in the backcountry, requiring passing across a couple of glaciers and a steep ridge or two separated by streams raging through a tangled North Cascades forest. A strong back and willing assistants are helpful. Once you get your rock to the lab, you must crush it to sand and isolate the zircon crystals from all the other minerals using magnets and liquids of varying densities. Hopefully there will be zircons, and enough of them. Zap the zircons with a beam of high-energy particles in a mass spectrometer, a laboratory instrument that measures the masses and relative concentrations of isotopes driven off by the beam. Compute the isotopic ratios, use a computer program to perform higher-math functions, and shazam! An age is obtained.

Oh, by the way, it could cost more than $1,000 per sample to do the spectrometry. Before you go through the effort and spend all that money, you want to be pretty sure that the target minerals in your sample rocks have not had their radioactive decay clocks reset by heating or chemical alteration. For instance, if a dike intruded your granite outcrop, the heat from the magma may have reset the decay clock of the zircons in the granite to a younger date. When the rock was in the crust, hot water might have chemically altered the minerals and made them unsuitable for analysis, too. This might mean that the date of a mineral is only a minimum age; however, this information can still be useful for understanding geologic history. After all that work, though, it is hard to top the feeling you get when a quality isotopic date supports your hypothesis about the age and origin of rocks you are studying.

Now you have the essential tools and geologic background to begin your exploration of western Washington's diverse geology. It's time for visits to the sites described in the vignettes. Happy travels, and have fun!

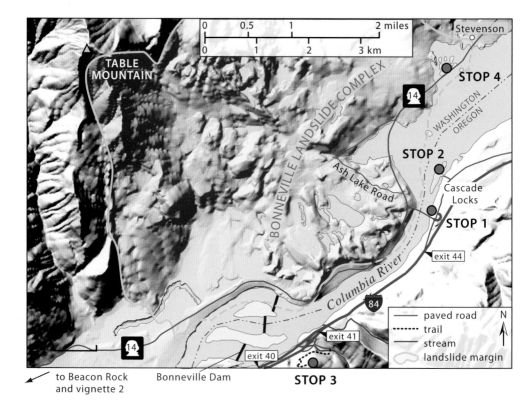

to Beacon Rock
and vignette 2

Bonneville Dam

STOP 3

GETTING THERE: From the I-205 junction in Vancouver, drive east on Washington 14 for 35 miles (56 km). Coming from the east, take Washington 14 to just west of Stevenson. Cross the Columbia River on the historic Bridge of the Gods toll bridge (constructed in 1926) into the town of Cascade Locks, Oregon. There is a one-way toll. Turn left onto Wa Na Pa Street (Cascades Locks Highway) and then make a quick right and park directly under the bridge at the southern concrete pier, stop 1.

To reach stop 2, turn left onto Wa Na Pa Street. Go 0.5 mile (0.8 km) north through the center of town and turn left into Cascade Locks Marine Park (watch for a sign for Sternwheeler Cruises). Veer left after the overpass, park at the old locks, and walk across the footbridge to Thunder Island.

To reach the Wauna Viewpoint Trail and stop 3, head through town on Wa Na Pa Street and get on I-84 west at exit 44. Drive 5 miles (8 km) and take exit 40 (Bonneville Dam); cross under the interstate and take I-84 east to exit 41, Eagle Creek (there is no westbound exit for Eagle Creek). At the bottom of the exit ramp turn right on NE Eagle Creek Lane and, after 0.2 mile (0.3 km), park at the picnic area just north of the Eagle Creek hikers' suspension bridge. Hike across the suspension bridge. To reach the viewpoint requires a 3.25-mile (5.2 km) round-trip hike on a moderate trail. A Northwest Forest Pass or day-use permit is required; day passes can be purchased at a kiosk in the first parking lot on the left after leaving the interstate off-ramp.

Stop 4 requires recrossing the toll bridge. Return to I-84 and head east. After 1.7 miles (2.7 km) take exit 44 and merge onto Cascade Locks Highway, which becomes Wa Na

1

Native Legend as Geologic Fact

BRIDGE OF THE GODS AND
THE BONNEVILLE LANDSLIDE

Native histories tell of a natural bridge that once crossed the Columbia River. There are many variations of the story. One recounts the legend of two warrior brothers, Wyeast (Mount Hood) and Pahto (Mount Adams), who both loved the fair maiden Loowit (Mount St. Helens). The warriors hurled burning rocks at each other and crossed the Columbia River on the Bridge of the Gods in order to fight for Loowit's affection. The fighting angered their father, Old Coyote, who collapsed the bridge to separate the unruly warriors. According to stories told to early settlers in the Columbia River region, Native Americans "not long ago" had been able to cross the river on a natural bridge. Fanciful illustrations grew out of these reports, including paintings of an elegant, arched natural bridge capped by a forest, gracefully soaring high over the turbulent waters of the river.

Geologists have shown that the legends are not purely fiction but reflect an actual event that took place a few generations prior to the arrival of American explorers in the Columbia Gorge. There really was a bridge of sorts across the mighty Columbia River, and Native Americans could, for a few months anyway, walk from bank to bank without getting wet feet. The "bridge" was actually a thick pile of rubble, the toe of a gigantic landslide that originated from the mountains north of the river, near present-day Bonneville Dam.

Pa Street. Cross under the Bridge of the Gods and immediately turn right to access the bridge. After crossing Bridge of the Gods, turn right (east) onto Washington 14. After 0.1 mile (0.2 km) turn left onto Ash Lake Road. Follow Ash Lake Road, which loops around and returns to Washington 14. Continue east 0.8 mile (1.3 km) on Washington 14. Just after you begin to drive along the shore of Rock Cove, watch for a road entrance to the north, across the highway; it leads into a turnaround area on a small "island" (stop 4) in the cove. If you miss it, no matter; there is an equally good substitute a mere 0.1 mile (0.2 km) farther east.

27

Features of the landslide are located on the Washington side of the Columbia River between Bonneville Dam and the town of Stevenson. The best overviews, however, are on the Oregon side, so our field trip begins in the town of Cascade Locks. Stop 1, at the south pier of the Bridge of the Gods toll bridge, showcases a mural by Portland-area artist Larry Kangas. The mural depicts a romanticized view of a graceful natural bridge over the river, providing an interesting juxtaposition to the actual bridge we will explore. The river is narrow here, only 0.2 mile (0.3 km) wide, pinned against the southern valley wall by the landslide.

Continue upriver through town to stop 2. Walk across the footbridge over the old river-steamer lock. The lock was built to help boats get above the steep drop of the Cascades Rapids. Be sure to look at the interpretive displays here, including a drawing of what the landslide dam looked like when it was fresh. (Note that the most recent geologic mapping conflicts with this interpretation; it does not appear that the landslide extended as far as Thunder Island or the Oregon side of the modern river channel.) The western edge

This mural, painted by Larry Kangas, depicts the legendary Bridge of the Gods.

of Thunder Island provides an overview of the geologic cataclysm disguised as mythic tale. The islands you see protruding out of the river are landslide blocks. The water is pretty deep here, so these are large blocks!

When the Corps of Discovery of Meriwether Lewis and William Clark descended the Columbia River about 3 miles (5 km) above this site in 1805, Clark marveled at "the trunks of many large pine trees s[t]anding erect as they grew at present in 30 feet [of] water," which he termed a "remarkable circumstance" in his journal entry for October 30. The snags he wrote of were found along the 40-mile (64 km) stretch of river between the modern sites of The Dalles, Oregon, and Stevenson. Clark, a highly competent adventurer but notoriously poor speller, wondered what could have raised the river to drown the trees. He wrote that "this part of the river resembles a pond partly dreaned leaving many Stumps bare both in & out of the water." He further noted that "the Stumps of pine trees . . . gives every appearance of the rivers damed up below from Some cause which I am not at this time acquainted with."

The very next day the expedition was perturbed to discover a serious obstacle in the river just below the modern site of Stevenson. Stop 2 overlooks the very same place, though much changed

321. WIND MOUNTAIN AND SUBMERGED FOREST, COLUMBIA RIVER.

Penny postcard, circa 1920, Card #321, Published by Chas. S. Lipschuetz Company, Portland, Oregon. —Courtesy of Lyn Topinka

by modern technology. Clark's journal described a 0.5-mile-long (0.8 km) "Great Shute" where the river fell 20 feet (6 m) and "the water of this great river Compressed within the Space of 150 paces in which there is great numbers of both large and Small rocks, water passing with great velocity forming [foaming?] & boiling in a most horriable manner." Once beyond this nasty bit of fast water, the river fell another 45 feet (14 m) in 2.5 miles (4 km) of continuous rapids. The intrepid explorers watched Native Americans portage the full length of this turbulent section of river, which extended downstream from the waterfront park, under the toll bridge, and around a curve to the site of Bonneville Dam. On November 1, Lewis and Clark and their men portaged all their gear and canoes around the Great Shute, a "bad way over rocks & on Slipery hill Sides."

The following day they avoided the next 1.5 miles (2.4 km) of rapids by portaging along the northern bank of the river, below the toll bridge, and ran the lower sets of rapids in the empty canoes. After passing a few more minor rapids downstream from where Bonneville Dam would be built, they reached the smooth lower reach of the Columbia at Beacon Rock (vignette 2), safely beyond the rambunctious portions of the Columbia River. The rocky narrows and rapids of the Great Shute allowed Clark to speculate on a cause for the drowned trees: "those obstructions together with the high Stones

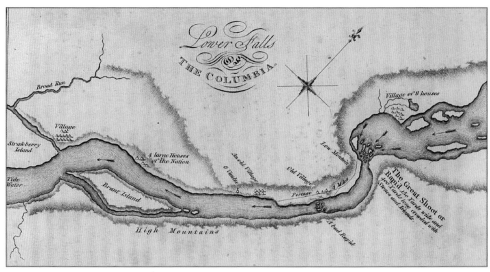

Captain William Clark's 1805 map notes that "the Great Shoot" is "crowded with Stones and Islands." The embayment on the north side of the river, just upstream of the rapids, marks the former course of the river before the Bonneville landslide buried a section of it. Bonneville Dam connects the three islands just below the rapids. —Courtesy of the Library of Congress

which are continually brakeing loose from the mountain on the . . . [north] Side and roleing down into the Shute aded to those which brake loose from those Islands above and lodge in the Shute, must be the Cause of the rivers daming up to Such a distance above."

The Great Shute the Corps of Discovery worked so hard to get around became known as the Cascades of the Columbia to the trappers of the Hudson's Bay Company post at Fort Vancouver, established a few years after the corps passed this way, and the mysterious drowned trees became "the submerged forest" of the Columbia. After struggling to get above the Cascades Rapids in 1825, Scottish explorer and botanist David Douglas named the entire mountain range for this obstacle. Both the Cascades Rapids and the trees became the subject of postcards and trips for tourists before the rapids and trees were submerged beneath the waters of Lake Bonneville in 1938, impounded behind the dam of the same name.

Clark's explanation for the origin of the drowned trees and the Cascades Rapids was mostly accurate. Alas, he did not describe a sufficiently catastrophic event for us to give him credit for recognizing that a single gigantic landslide had dammed the river. Another

Aerial view looking downstream at the Cascades Rapids and the brand-new Bridge of the Gods, circa 1926. The rapids, Lewis and Clark's Great Shute, form a distinct white V at the center. The rocky islets are landslide blocks. Thunder Island is the large island on the left. —Courtesy of US Army Corps of Engineers

century passed before that was fully recognized. In the meantime, other explanations were brought forth, invoking a variety of downward faulting or subsidence scenarios to explain the drowned trees, including collapse of the forested surface into an underground river channel and landslides carrying the trees themselves into the river. Some theorists proposed that dams drowned the forest, citing crustal uplift, volcanic action, and glacial moraines as the forces blocking the river. Missionaries Daniel Lee and Joseph Frost suggested a landslide dam as early as 1844, but they did not indicate a source for the landslide.

Finally, Ira A. Williams of the Oregon Bureau of Mines and Geology made a serious investigation of the geology in the area, which he published in 1916. His study determined that "gigantic landslides" from Table Mountain to the north had dammed the river. According to Williams, the most recent of these giant slides, now called the Bonneville landslide, rerouted the Columbia 1 mile (1.6 km) to the south; had not been completely eroded away; and, until the river rose to overtop it, had been high enough for native inhabitants to walk across.

The steep south face of Table Mountain is the scarp left by the Bonneville landslide, which underlies the forested terrain just across the Columbia River. Stacked lava flows of the Columbia River Basalt Group show clearly. The light-colored rock scar to the right is the scarp left by the 2007 Greenleaf landslide. —Courtesy of Lyn Topinka

The Bonneville landslide is the youngest in the Cascades land-slide complex, a group of four large overlapping landslides. They slid, at different times, off the mountain walls on the Washington side of the Columbia Gorge between Bonneville Dam and the town of Stevenson. The landslides' source is marked by a prominent long rock scarp called Red Bluffs, running along the southeast flanks of Greenleaf Peak (elevation 3,424 feet, 1,043 m) and Table Mountain (3,417 feet, 1,041 m) and the ridge connecting them. The scarp is a near vertical rock wall up to 800 feet (245 m) high. Dense lava of the Columbia River Basalt Group, seen in the cliffs of the two peaks, lies on top of weaker volcanic breccia and sediment of the Eagle Creek Formation, which is rich in clay minerals. The rock layers all dip slightly toward the river, the result of folding during uplift of the Cascade Range. Erosion more easily undercuts the softer Eagle Creek rocks, leaving the dense lava lying high above the river val-ley. Water percolating into the Eagle Creek rocks lubricates the clay under the thick basalt, and gravity can do the rest. The resulting landslide deposits are disordered, lumpy terrain extending for about 5.5 miles (8.9 km) along the north bank of the Columbia, cov-ering about 13 square miles (34 km^2).

The Bonneville slide, the largest in the complex, covers 5.5 square miles (14.3 km^2). It slid southeast about 4 miles (6.4 km) from high on the south flank of Table Mountain and buried and dammed the river. The landslide rubble is a very poorly sorted deposit that includes everything from fine sand to rock slabs 200 feet (60 m) thick and 800 feet (245 m) long. The deposit extends for 2.5 miles (4 km) northeast to southwest along the valley floor, from Ashes Lake to Bonneville Dam, and is 300 to 400 feet (90 to 120 m) thick in places. The edge of the landslide farthest from the source lies beneath Lake Bonneville. Many small lakes fill depressions on the landslide's toe, caused by a drainage disordered by the hummocky terrain that is a signature of landslide deposits.

After the landslide, the Columbia River began to back up. The size of this temporary lake has been estimated based on the height of the surviving landslide deposit and the elevations at the top of a number of deltas that tributaries quickly built out into the lake. The lake was 275 to 300 feet (84 to 90 m) deep at the landslide dam and extended up the Columbia Gorge 160 miles (260 km), as far as Wal-lula Gap south of Pasco. The surface elevation of this lake, dubbed the Lake of the Gods, was as high as 300 feet (90 m) above sea level. For comparison, Lake Bonneville, behind the eponymous dam, has a surface elevation of 72 feet (22 m). The Lake of the Gods was twice as long as, and far deeper than, the modern lake, with more than

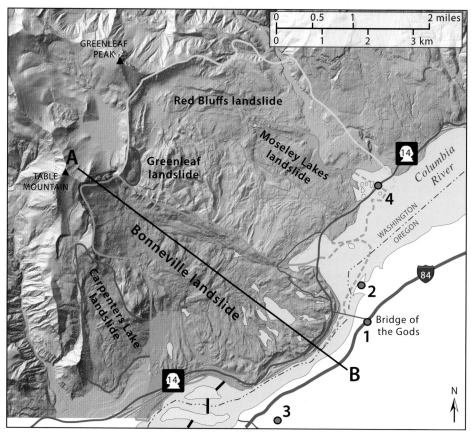

Boundaries of the Cascades landslide complex. The prelandslide course of the river was somewhere to the west of the current course, beneath the toe of the Bonneville landslide (orange outline). Black line indicates cross section below. (Topography generated from lidar data.)

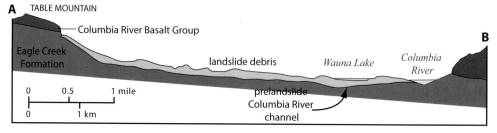

This simplified northwest-southeast cross section shows the dipping contact between the Columbia River Basalt Group and the older Eagle Creek Formation. This contact enhances the potential for sliding due to gravity and water permeation. (Adapted from Norman and Roloff 2004.)

one hundred times the volume. The landslide dam thrown across the Columbia River would indeed have allowed Native Americans to cross the river on dry ground.

At today's typical flow rates, it would only have taken a few months for the Columbia River to fill the lake. The impounded water eventually overtopped the lowest part of the landslide's toe and began to reestablish a new channel 1 mile (1.6 km) south of its former course. The jury is out as to whether the dam was breached catastrophically, gradually, or episodically. One interpretation posits that a deposit of large boulders just downstream from Bonneville Dam represents a very large flood, which swept down the river when the river initially breached the landslide dam. The initial dam failure probably was a catastrophic event as the reinvigorated river rapidly stripped away loose landslide debris, but enough material was left to back up the Columbia, albeit at a lower level, until 1938. Persistence of the Cascades Rapids shows that the river was still cutting downward through the landslide. When Bonneville Dam's floodgates were closed in 1938, the lowest point cut through the landslide dam was 45 feet (14 m) above sea level, the river level at low water. The filling of Lake Bonneville finally drowned the Cascades Rapids and the submerged forests, putting an end to the river's erosion of the remaining landslide deposit.

For a view across the river to the landslide's toe and source area, make the hike to stop 3 on the Wauna Viewpoint Trail. After crossing the Eagle Creek suspension bridge, turn right. Be aware that poison oak is found along this trail. Follow the wide, moderate trail in old-growth forest through a few switchbacks to the junction with the Gorge Trail. Take a sharp left here and hike upward through six more switchbacks to the trail's end beneath a power line tower (elevation 1,075 feet, 328 m).

The hummocky surface of the Bonneville landslide and the source area on Table Mountain are quite visible across the river. From here you can visualize how the landslide crossed the river valley and impinged upon the southern valley wall just below. In the distance, the spidery Bridge of the Gods marks the bottom of Clark's Great Shute. The river makes a gentle sweeping curve between the bridge and Bonneville Dam, marking the landslide's toe. The former river course passed under the landslide surface about 1 mile (1.6 km) north of today's drowned channel. The Lake of the Gods overtopped the low point of the landslide at its toe to cut a new channel. The Corps of Discovery made their arduous canoe portage of the lower rapids along the north shore of the river. The roaring, bustling interstate, railroad, and river-barge routes pass through the narrow corridor between the landslide and the south wall of the gorge. The confined transportation corridor through here should remind us that critical transportation infrastructure is vulnerable to future landslides. The high-tension power line right-of-way beginning at Bonneville Dam is intentionally located on the Oregon side of the river to avoid this hazard.

The toe of the Bonneville landslide displaced the river to the south. Stop 3, Wauna Viewpoint, is just out of the photo, bottom right. —Courtesy of John Scurlock

To see landslide features up close, head to stop 4. On the way, take note of the rock blocks and deep hollows along Ash Lake Road; many hollows are filled with ponds, marking the chaotic surface of the 550-year-old landslide. In 1 mile (1.6 km), at the junction with Blue Lake Road, turn right and proceed along the north shore of Ashes Lake. You just left the surface of the Bonneville landslide. Though it is now isolated from the Columbia River by the highway causeway, Ashes Lake fills the abandoned course used by the Columbia prior to the landslide. You are now driving on the surface of the older, still undated Red Bluffs landslide.

Once you reach WA 14, turn left (east) and continue 0.8 mile (1.3 km) to stop 4. The small dome-shaped islands in Rock Cove are hummocks left behind by the undated Moseley Lakes landslide, the oldest of the four landslides composing the Cascades landslide complex. The water of Lake Bonneville accentuates the hummocks, surrounding them and covering the lower lumpy topography. Hummocky terrain is characteristic of landslide deposits. It forms when

The landslide hummocks in Rock Cove are remnants of the Moseley Lakes landslide.

the chaotic, churning landslide surface comes to rest. The slide doesn't stop all at once; more mobile portions may pile up against large blocks that have already stopped. Erosion tends to round these landforms

The timing of the landslide has been a controversy, with much conflicting evidence obtained from tree rings, wood samples (radiocarbon dating), and even lichen, which can be dated based on known growth rates. The fresh appearance of the drowned trees convinced Lewis and Clark that the river must have been dammed only a few decades before their expedition. The Native American oral histories were widely interpreted as evidence for a very young age, as were the still-standing drowned trees. But early radiocarbon dates obtained from buried wood were around 1100 CE; more modern analysis of these samples indicates the trees died in the first half of the fifteenth century. The downstream flood deposits used as evidence for catastrophic breaching of the dam are covered with ash that erupted from Mount St. Helens between 1479 and 1482 CE.

The convoluted process of determining the age of the slide is a good reminder of how the scientific process works. Many stones are

The 2007 Greenleaf landslide scarp on the Red Bluffs. Springs issue from the upper surface of a clay-rich layer within the Eagle Creek Formation, attesting to the amount of lubrication that can develop within rock layers. Light-colored lava flows and intrusions alternate with unconsolidated volcanic deposits above. The south face of Table Mountain, left, reveals much thicker Columbia River Basalt Group on top of Eagle Creek rocks. The trees on the skyline are 60 to 80 feet (18 to 24 m) tall. —Courtesy of Don Nelson

turned over, and peers must review scientific conclusions. Conflicting dates for the same event are clues that more work needs to be done. The current age interpretation of the Bonneville landslide is the middle of the 1400s, just before the Mount St. Helens ash fell.

We should not think of these giant landslides only in terms of the past. Relatively smaller landslides still occur in the Cascade landslide complex. The Greenleaf slide occurred in December 2007 when a section of the Bonneville landslide scarp at the south edge of the Red Bluffs slid about 1,000 vertical feet (300 m) and 0.5 (0.8 km) out from its source, burying some of the Bonneville rubble. The slide debris covers about 38 acres (15 ha). Smaller slides are spawned by the same factors as the larger, ancient ones: (1) water percolates downward through thick layers of the Columbia River Basalt Group and permeable upper strata in the Eagle Creek Formation; (2) the water accumulates and lubricates the top of a relatively impermeable clay-rich rock layer lower in the Eagle Creek; (3) the entire stratigraphic package dips toward the river; and (4), gravity.

The small slides should serve as reminders that the threat of huge landslides persists. If an event the magnitude of the Bonneville landslide happened today, there is potential for great economic loss and human tragedy. The narrow transportation corridor in the Columbia Gorge is vulnerable; shipping lanes, highways, electric power lines, and gas pipelines all lie along the bottom of the valley. The catastrophe inherent in the overtopping or destruction of Bonneville Dam, either directly from a landslide or by a wave generated by a slide entering the lake, is almost too horrible to contemplate, not to mention the specter of the subsequent gigantic flood as Lake Bonneville drains. It is unlikely that such a huge landslide could be predicted, and it surely cannot be prevented, so learning to live with the potential for such low-frequency, but very high consequence, events is the only option.

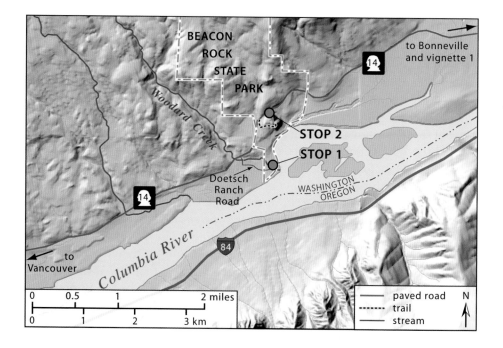

BEACON ROCK STATE PARK

Woodard Creek

STOP 2

STOP 1

Doetsch Ranch Road

WASHINGTON
OREGON

14

14

14

to Bonneville
and vignette 1

to Vancouver

Columbia River

84

| 0 | 0.5 | 1 | 2 miles |
| 0 | 1 | 2 | 3 km |

——— paved road N
------- trail
——— stream

GETTING THERE: Beacon Rock towers directly above Washington 14 in the Columbia Gorge. A good place to view the entire mass without craning your neck is stop 1, a boat launch. From the junction of I-205 in Vancouver, Washington, head east on Washington 14 about 27.5 miles (44 km) and turn right on Doetsch Ranch Road. Follow signs to the boat launch, 0.6 mile (1 km) down the road. The turnoff to Doetsch Ranch Road is about 5 miles (8 km) west of Bonneville Dam.

Head back to Washington 14 and turn right. The trailhead parking lot for Beacon Rock, stop 2, is a less than 1 mile (1.6 km) down the road, on the right. The trailhead is just beyond the west edge of the parking area. The trail is open year-round, 8 a.m. to dusk. The spectacular but easy hike is 2 miles (3.2 km) round-trip and gains 650 feet (200 m) of elevation. Beacon Rock is a Washington State Park, so a Discover Pass is required.

2

The Rocky Guts of a Volcano

BEACON ROCK AND THE MISSOULA FLOODS

Enormous Beacon Rock, an isolated, pointed, 850-foot-high (260 m) pillar, rises straight up from the bank of the Columbia River and is visible for 20 miles (30 km) up and down the river. Meriwether Lewis, William Clark, and the Corps of Discovery paddled beneath this landmark during their quest for the Pacific Ocean. On October 31, 1805, as his men negotiated the rapids caused by the Bonneville landslide (vignette 1), Captain Clark provided the first written description of the high promontory: "a remarkable high detached rock Stands . . . about 800 feet high and 400 paces around, we call the 'Beaten rock.'" A later diary revision referred to Beacon Rock, and this is probably what he originally meant. The following April, Meriwether Lewis, who couldn't spell much better than Clark, noted: "This remarkable rock which stands on the North shore of the river . . . rises to the hight of seven hundred feet; it has some pine or reather fir timber on it's northern side, the southern is a precipice of it's whole hight. it rises to a very sharp point and is visible for 20 miles below on the river."

At stop 1 you can see the high south face of Beacon Rock rising from the river. It is a plug of lava that once filled the interior of a cinder cone, a type of volcano. The plug belongs to one of the youngest volcanoes—and apparently the farthest east—of the Boring Volcanic Field, which covers a wide area, including the Portland-Vancouver metropolitan area. Mount Tabor Park in Portland includes a 203,000-year-old cinder cone of this volcanic field. Uneroded cinder cones may be recognizable by their classic shape: a truncated cone. They are made from cinders, which are blobs of lava erupted into the air that cool, fall, and accumulate in a heap around the eruption vent to form a cone. The rim of the crater forms the flattened summit profile. Cinder cones may persist for hundreds of thousands of years in dry, unglaciated climates. Pristine cinder cones include Lava Butte near Bend, Oregon; Sunset Crater near Flagstaff, Arizona; and Paricutín in central Mexico. The latter is the famous "volcano that grew in a corn patch" beginning in 1943.

41

There are many cinder cones on the west flank of Washington's Cascade Range; unfortunately all are forested and difficult to visualize as cones. You won't, however, find a pile of cinders at Beacon Rock. They were stripped away to reveal the massive lava plug. If there was ever a lava flow associated with Beacon Rock, it is gone too.

Beacon Rock towers over the Columbia River at stop 1.

Cinder Cone at Lassen Volcanic National Park in California shows the classic form. Though lava flows spread from the base of the cone and dam a creek to form a lake (bottom), the cone itself is composed of cinders.
— Courtesy of John Scurlock

Proceed to stop 2. At 265 feet (80 m) of elevation, you are about 0.25 mile (0.4 km) of the way to the top of the plug. The eruption that produced the Beacon Rock volcano began when magma rich in pressurized gases, such as sulfur dioxide, steam, and carbon dioxide, rose to the surface, where it sprayed forth a few hundred feet into the air as frothy molten lava. This cooled and hardened as cinders, which built a growing pile near the volcano's vent and formed a cinder cone. Geologists also refer to cinders as scoria. Scoria is a lacy looking volcanic rock that has about equal parts rock and vesicles, the holes that formed as gas escaped the cooling cinders. As long as the magma contained plenty of gas, the young volcano erupted explosively, and scoria piled up around the vent. Pressure differences between the magma reservoir and the surface drove the magma upward through underlying rocks and the growing cone as an incandescent, pasty mass, where it cooled as today's plug.

The Beacon Rock volcano reached skyward far above the top of the plug and extended outward radially many times wider than the base of Beacon Rock. The outer portion of the original lava plug intermingled with the cinders it was invading. The vertical walls of Beacon Rock we see today, however, are the eroded remnants of the original plug, which was emplaced below the surface. The temperature of the plug gradually decreased from around 1,800°F (1,000°C) to the ambient air temperature. As the rock cooled, it contracted. The loss of volume due to contraction was taken up by fracturing. The fractures progressed inward as the mass of lava cooled, from the top down and from the sides inward, and eventually intersected, defining columns. The columns can be hexagonal or have as few as three or as many as twelve sides. Though columns such as these are often called columnar basalt, columns can form in cooling magma of any composition (see also vignette 22). There are endless miles of basalt columns in eastern Washington, and andesite columns can be seen along roadsides at Mount Rainier and Mount Baker. Beacon Rock consists of basaltic andesite, midway between basalt and andesite in composition.

In the rock face above the eastern parking lot, the columns are on their sides, exposed in cross section as multisided polygons. These fractures extend into the plug. Since cooling fractures migrate into the hotter interior of a body of lava from the cooler outer part, the eastern columns must be near the original outer edge of the plug.

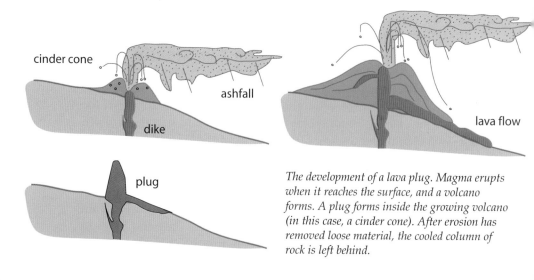

The development of a lava plug. Magma erupts when it reaches the surface, and a volcano forms. A plug forms inside the growing volcano (in this case, a cinder cone). After erosion has removed loose material, the cooled column of rock is left behind.

The different orientation of the columns on different sides of Beacon Rock demonstrate that cooling in large masses of rock can progress inward in many directions. We can surmise that the outer margin of the plug on the south and west sides initially cooled inward, too, but the outer, horizontal columns have since eroded away, exposing the vertical columns of the plug's interior. These vertical columns are best viewed from stop 1.

Beacon Rock has been called one of the largest monoliths in the world; some Internet sources say it is second only to the Rock of Gibraltar or, alternatively, second to Devils Tower in Wyoming. Monolith means "single rock," and in general, dictionaries define the term as a single block of stone unbroken by fractures. Strictly speaking, this means that Beacon Rock is not a monolith, nor is

The horizontal orientation of the columns on the east side of Beacon Rock suggests this side of the plug has not eroded much from its original margin.

Devils Tower. There is plenty of debate over what the world's largest true monolith is, or how such a geographic feature can even be measured. The list includes Uluru (Ayers Rock) in Australia; El Capitan in California; Sugarloaf in Rio de Janeiro; and yes, the Rock of Gibraltar. But alas, it should not include Beacon Rock.

The human history of this volcanic plug is as interesting as the geology. According to the diaries of Lewis and Clark, the tribes who traded on the Columbia recognized this place as the end of difficult paddling when westbound, and as the uppermost reach of tides; Native Americans reportedly called it Che-che-optin, reputed to mean "navel of the world." Around 1900 the US Army Corps of Engineers proposed blasting the rock to provide material for jetties at the mouth of the Columbia. Owners of the railroad at its base objected, as flying rock fragments could pose a serious hazard to trains. Because the rock is heavily fractured, rock quarries were also proposed.

Frank Smith, Charles Church, and George Purser made the first ascent of Beacon Rock in 1901. They left behind iron pitons and fixed ropes for the use of others. In 1914, forty-seven members of Portland's famed mountaineering club the Mazamas repeated the climb. If only they had waited four more years.

Enter one Henry Biddle, an engineer from Philadelphia, who purchased the entire mass of rock in 1915 from Charles E. Ladd for $1. Ladd, a Portland banker opposed to the destruction of Lewis and Clark's landmark, sold it to Biddle with the understanding that he would work to preserve it. Admittedly, a dollar went further in those days, but this still seems like a bargain. Biddle had a monomaniac's dream: to build a 1-mile (1.6 km) trail to the summit, as Biddle himself put it, "in perhaps the most difficult location in which a trail had ever been built," to serve as a tourist attraction. A crew began building the eye-popping trail in 1915, blasting into the rock to make way for fifty-two switchbacks, with cable for safety rails, and twenty-two wooden bridges to span blank rock faces. The crews completed the fantastic trail in 1918, at a cost of $15,000, and it immediately became a popular weekend destination for Portland residents, without charge. Trailside plaques commemorate this remarkable achievement. The wooden bridges are now replaced with steel, and the cable safety rails with stouter metal guardrails.

At Biddle's death, the terms of his estate required that the rock and surrounding land be offered for $1 to the state of Washington as a park. At first the state refused, not wanting to incur the responsibility, but Washington agreed to the terms in 1935 to prevent Oregon from purchasing the parcel.

The switchbacks up Beacon Rock.

A hike up Beacon Rock is de rigueur for visitors. The spectacular trail uses Biddle's route up the west face. There is plenty of vertical exposure, but a sturdy handrail runs the entire length. Most of the trail can be walked in street shoes. The Columbia Gorge is famous for its wind, so be prepared. There are interpretive signs describing the geology at Beacon Rock, written by the Ice Age Floods Institute.

Before you begin your hike, pick up a piece of rock in the talus and carry it along until you get into good light. Weathering and lichens quickly obscure surface features of the rock, so find a freshly broken surface. This is a good time to use a hand lens, as the minerals in this rock are very tiny. The rock has a fine-grained texture, a term referring to the relative size, frequency, and orientation of mineral crystals and structures, such as vesicles. The size, type, and relative proportions of crystals in lava depend on the chemistry and cooling history of the magma. The lighter-colored minerals in the rock are plagioclase, a felsic mineral in the feldspar family that is relatively rich in silica; it precipitates early from the ionic stew in a magma chamber. While the magma is hotter, it is less viscous, so ions can migrate through the magma and link up to form minerals. Plagioclase continues to form for nearly as long as the magma is cooling; therefore, it has a chance to grow for a longer time and become larger.

The darker minerals are mafic, meaning they contain iron or magnesium in addition to the same elements found in plagioclase. Iron and magnesium are typical in magmas of the type that created Beacon Rock, but less abundant than the silica and aluminum in plagioclase. Mafic crystals are generally fewer and smaller due to these low proportions. Crystals can grow larger if magma—or its counterpart at Earth's surface, lava—has more time to cool. Crystal growth is halted when magma or lava cools to the point that it is too viscous for ions to migrate.

Beacon Rock lava is rich in crystals, but few are longer than 0.04 inch (1 mm) in length. The most prominent dark mineral is orthopyroxene, which looks like bits of brown glass. These are dispersed between clear to white plagioclase grains, the most common mineral in this or any lava in the Cascades. The plagioclase crystals form a sugary groundmass, but with good light you'll see that the larger ones are clear needles. There are also many tiny black crystals; these are clinopyroxene. The fact that this rock is mostly crystals indicates that the mass of lava in the cone cooled slowly. If it had cooled quickly, it would have formed glass, which doesn't have the orderly atomic arrangement that gives gem-quality crystals their interesting shapes. Beacon Rock's lava is about 55 percent silica and can be pigeonholed as basaltic andesite.

There are almost constant views of the Columbia River from the cleverly constructed trail; the best are just below the top. The rock is noticeably redder around the summit due to oxidation. This is good evidence that we are near the original top of the plug, where the hot rock oxidized through contact with oxygen and probably water percolating downward through the cinder pile as it cooled. Watch very carefully for a single tiny exposure of loose red scoria preserved in tree roots on the last switchback before the top; these are about all that remain of the extensive scoria blanket that once formed the cone.

So what stripped away the outer jacket of cinders, and even a portion of the solid rock on the outer margins of the plug? We are south of the reach of the landscape-altering glaciers of Pleistocene time. A geologist faced with this problem might invoke the Columbia River, flowing right at the rock's doorstep, as the agent of erosion. In fact, figuring this to be the case, geologists long assumed Beacon Rock was hundreds of thousands to millions of years old. With the obvious fact that the river has cut the Columbia Gorge through hundreds of feet of much older Columbia River Basalt Group, there seemed to be no mystery.

The view upriver from the summit. The Bonneville Dam complex lies at the foot of the rumpled terrain of the Bonneville landslide in the distance.

But the application of modern age-measurement techniques determined the plug's true age. Beacon Rock is a surprisingly young 57,000 years old. For the Columbia River to cut through the exposed 850-foot (260 m) thickness of basaltic andesite in that period of time would require an erosion rate of a whopping 0.2 inch (5 mm) per year. Lest this rate seem piddling, compare it with recent studies of erosion rates in the rainy western Cascades region: 0.008 to 0.01 inch (0.21 to 0.33 mm) per year. So, something out of the ordinary happened at Beacon Rock.

That something was the Missoula floods, among the greatest floods known in Earth's geologic record. These monstrous outpourings of water occurred at least eighty-nine times (some geologists say one hundred) between 20,000 and 15,000 years ago. When lobes of a continental ice sheet moved south out of Canada into northern Washington, Idaho, and Montana, one of the fingerlike lobes dammed the Clark Fork River near the Idaho-Montana border, which backed up to form Glacial Lake Missoula. This lake was up to 2,000 feet (610 m) deep and had a volume of as much as 500 cubic miles (2,080 km³)—about half the volume of Lake Michigan. Eventually, the ice dam failed, catastrophically releasing the lake's contents across eastern Washington.

The huge lake drained in as few as two days. The largest floods raced southwest across the land at speeds of 75 miles (120 km) per hour, with flow rates of up to 440 billion gallons (1.7 trillion liters) per second—ten times the flow of all the world's rivers combined! The floods scoured deeply into bedrock, carving and deepening Grand Coulee, Moses Coulee, and the complex of canyons of the Channeled Scablands, a landscape that owes its origin to these titanic floods. Laden with enormous volumes of rock and soil, the floods rocketed out of the basins of eastern Washington and drained via the Columbia into the Pacific Ocean. Once Glacial Lake Missoula had drained, the Clark Fork lobe of the glacier could advance again, blocking the river, and the process would repeat itself.

If there were one hundred Missoula floods, then a flood occurred, on average, every fifty years. The largest floods swept up against Beacon Rock, bursting out of the Columbia Gorge and entering the Portland basin with a surge of water 400 feet (120 m) deep and moving 20 miles (32 km) per hour or faster. At The Dalles, Oregon,

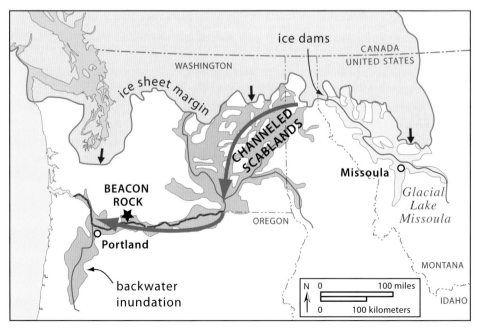

The Missoula floods scoured the Channeled Scablands before rocketing into the Columbia Gorge. Brown shading indicates areas inundated by the floods.

55 miles (88 km) upriver from Beacon Rock, soil was stripped and basalt bedrock scoured 1,000 feet (300 m) above the modern river. A recent study of flood heights shows that soil below 755 feet (230 m) of elevation was removed by the floods at Crown Point, 14 miles (22 km) downstream of Beacon Rock on the Oregon side. The cinders in the tree roots near the summit are good evidence that the Beacon Rock plug was not completely inundated. Based on data from elsewhere in the Columbia Gorge, it is safe to say that the water was certainly very high, nearly to the top of Beacon Rock.

The loose pile of cinders didn't stand a chance in the face of repeated onslaughts of sediment-laden water. Probably the first few floods removed the cinders blanketing the volcano's core, either by submerging them or stripping away the lower cindery skirts and allowing gravity to remove the rest. Subsequent floods finished the job, plucking away loose, fractured rock and abrading corners on the remaining plug. Some of the material removed from Beacon Rock was dumped in the Portland area, where up to 230 feet (70 m) of Missoula-flood sediment was deposited, including boulders 15 feet (4.5 m) across, but most of it was probably sluiced out into the Pacific Ocean, where a huge underwater fan is found off the mouth of the Columbia River.

The Troutdale Formation, across the highway from the Beacon Rock trailhead, comprises coarse alluvial deposits of the Columbia River that predate Beacon Rock.

Back at the trailhead, look at the bank of stratified boulders and sand in the roadcut across the highway. This sediment belongs to the Troutdale Formation of late Miocene and Pliocene age. The ancestral Columbia River deposited the formation prior to any eruption in the Boring Volcanic Field. It was deposited at about the same elevation as the modern Columbia River, which is essentially at sea level here, and subsequently raised to this elevation as the Cascade Range was uplifted. The Troutdale Formation here was invaded by the Beacon Rock magma plug and also later stripped by the Missoula floods. Its presence this close to the plug tells us that erosion has not removed much of the east side of the plug.

3

Changes at the Mouth of the Columbia River

CAPE DISAPPOINTMENT

People have lived along the Columbia River for thousands of years. Wimahl ("big river" in the Chinookan language) has been the economic and cultural wellspring for the dozens of Native American tribes who settled its banks. Teeming with fish and birds, it provided people with a greater larder than they could possibly exploit.

An armada of distinguished European navigators failed to discover the river. On August 17, 1775, Bruno de Hezeta mistook the river's mouth for the Strait of Juan de Fuca. Unable to cross the treacherous bar of sediment at the Columbia's mouth, Hezeta made a map of what we now know to be an estuary, and he named a high promontory on the north shore Cabo de San Roque. Though he did not recognize it as a river, Rio de San Roque began appearing on nautical charts soon after. In 1778 the venerable James Cook missed the river's mouth altogether. In 1788 English fur trader John Meares, searching for the Rio de San Roque, declared that "no such river . . . exists," and he renamed Hezeta's promontory Cape Disappointment to reflect his mood at this "discovery." George Vancouver, sailing in the *Discovery* in April 1792, sighted Cape Disappointment and noted the change in the color of the water offshore, but he didn't consider the "opening worthy of more attention" and thus sailed northward.

Discovery of the river was left to the American fur trader Robert Gray aboard his ship *Columbia Rediviva*. A few days after Vancouver continued north up the coast from Cape Disappointment, he encountered Gray, who promptly set sail for the place. Gray and his men crossed the seething bar at the river's mouth on May 11, 1792, and entered the estuary. He named the great river after his ship and made a chart of the lower 12 to 15 miles (19 to 24 km) or so. The Lewis and Clark Expedition reached the Pacific Ocean at Cape Disappointment on November 18, 1805. Since then, easily visible geomorphic changes have occurred near the cape.

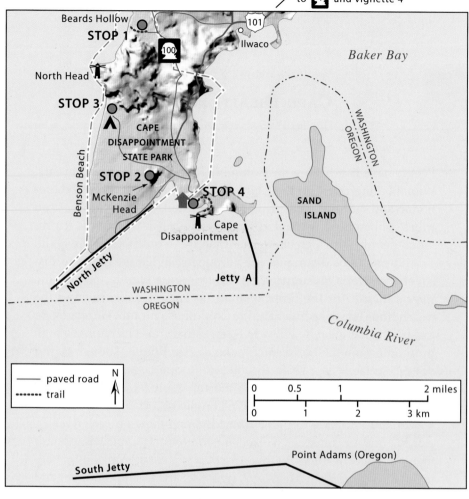

to 4 and vignette 4

Beards Hollow

STOP 1

US 101

Ilwaco

Baker Bay

100

North Head

STOP 3

WASHINGTON
OREGON

**CAPE
DISAPPOINTMENT
STATE PARK**

STOP 2

McKenzie
Head

**SAND
ISLAND**

STOP 4

Cape
Disappointment

Benson Beach

North Jetty

Jetty A

WASHINGTON
OREGON

Columbia River

— paved road

····· trail

N

| 0 | 0.5 | 1 | 2 miles |
| 0 | 1 | 2 | 3 km |

Point Adams (Oregon)

South Jetty

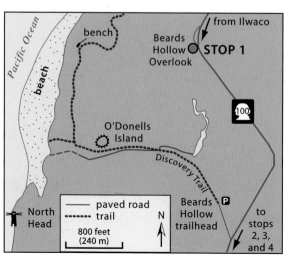

Pacific Ocean

bench

from Ilwaco

Beards
Hollow **STOP 1**
Overlook

beach

100

O'Donells
Island

Discovery Trail

North
Head

Beards
Hollow
trailhead

P

to
stops
2, 3,
and 4

— paved road

····· trail

N

800 feet
(240 m)

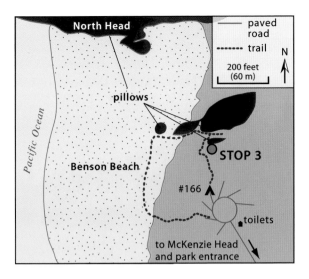

GETTING THERE: Cape Disappointment State Park is at the extreme southwestern tip of Washington, just beyond the town of Ilwaco off US 101. Depending on where you are in the state, there are multiple routes that lead to US 101, including Washington 4, which parallels the Columbia River from I-5 at Kelso, and Washington 101 along the coast.

Set your odometer to 0 in Ilwaco where US 101 abruptly turns north toward Long Beach. The distances to the stops are from this intersection. Follow the signs for Washington 100 Loop (North Head Road). Stop 1, Beards Hollow Overlook, is a viewpoint with parking at 1.3 miles (2.1 km).

For stops 2 through 4, continue south on Washington 100. At 2.9 miles (4.7 km) turn right at the junction with Washington 100 Spur. After 3.4 miles (5.5 km) take a right at the park entrance (pick up a park campground map here) and turn right again toward McKenzie Head (stop 2). At 4 miles (6.4 km) park in the parking lot on the south side of the road.

To reach stop 3, a cluster of rock knobs consisting of pillow lavas, continue beyond McKenzie Head toward the ocean, following signs to the campground. Aim for the campground's north end; you are looking for the cluster of sites numbered 161–170. Park by the toilets at this group of camps at 4.9 miles (7.9 km). If site 166 or 167 is unoccupied, walk through it to the high rocky outcrops 30 yards (30 m) to the north. If these sites are occupied, walk a few yards on the track between site 167 and 168 to the beach and head north. To get to the pillows behind site 166, head inland at the first rock outcrop in the sand; you'll need to pick your way over piled driftwood and through some small trees to get to the exposures.

Stop 4 is the Lewis and Clark Interpretive Center. Return to Washington 100 Spur, turn right, and drive 0.5 (0.8 km) to the center, which offers expansive views of the Columbia's mouth and the wide Pacific. Contact the center before your visit for current hours and fees.

Cape Disappointment State Park is the most heavily used park in the Washington State Park system, so be prepared for company, particularly in the summer. Please remember that rock hammers cannot be used in the park, and collecting of any kind is not allowed. A Discover Pass is required to visit the park's trailheads.

Cape Disappointment, underlain by Eocene-age pillow lava and breccia of the Crescent Formation, juts into the Pacific at the mouth of the Columbia River and is topped by a 53-foot-tall (16 m) lighthouse.

The Columbia River discharges more water into the eastern Pacific Ocean than any other; an average of 275,000 cubic feet (8,000 m³) passes through the river's 3-mile-wide (4.8 km) mouth each second. The Columbia is the third-largest river in the United States by discharge and the largest outside the Mississippi-Missouri system. The drainage basin, 258,000 square miles (670,000 km²), includes all of eastern Washington, northern and eastern Oregon, most of Idaho, the southeast corner of British Columbia, northwest Montana, and small parts of Nevada, Utah, and Wyoming.

The Columbia's huge drainage basin supplies a tremendous sediment load to the river. One study estimates that since the start of the Holocene period, 30 cubic miles (130km³) of sediment has accumulated at the mouth of the Columbia. A large volume of this sediment must be related to the humongous Missoula floods that swept across eastern Washington during the Pleistocene. In modern times, prior to the construction of more than two hundred dams, the river delivered 11.3 million cubic yards (8.7 million m³) of sediment per year, nearly twice the volume of the old Kingdome in Seattle, to the estuary just above the river's mouth; nearly half of the sediment load was sand. Since all those dams were built the load has decreased by about half, to 5.6 million cubic yards (4.3 million m³). Sand accounts for just under a third of this. Though the dams trap most of the sediment now, there is still a considerable amount reaching the Pacific. Major tributaries entering the Columbia below the lowest dam at Bonneville supply a lot, but there is also a large amount of sand in the riverbed from pre-dam days.

As the river reaches the ocean its current slows. Its capacity to carry sediment decreases, and coarser sediment drops out. Sand and silt accumulate as the famous Columbia Bar, a subtidal delta that is among the most fearsome mariners' graveyards in the Pacific. More

The Columbia River's drainage area. (Wikipedia, Kmusser – self-made, based on USGS and Digital Chart of the World data.)

than two thousand large vessels have foundered here since early in the nineteenth century, claiming more than 1,200 lives. Countless other small boats have been sunk or crushed by waves breaking over the shallow sands of the bar.

Powerful currents sweep along the coast. These flow strongly to the north in the winter and weakly toward the south in summer. The currents sweep much of the sediment away from the river mouth and deposit it along the coast to the north and south. Columbia River sediment is carried as far north as Point Grenville, 72 miles (115 km) up the coast of Washington. The Long Beach Peninsula, just north of the Columbia, is a 22-mile-long (36 km) spit of Columbia River sand built across the mouth of Willapa Bay by longshore drift (see vignette 17 for more about spit formation).

Humans have had a major influence on the location of the shoreline on this stretch of coast. The long jetties extending seaward from Cape Disappointment and Point Adams (in Oregon) were built between

A model showing coastal sediment dispersal from the mouth of the Columbia. Black arrows show direction and the relative volume of sediment transported away from the river. Two-headed arrows indicate seasonal changes in the direction of longshore drift.

1885 and 1917. The jetties are designed to rise 28 to 32 feet (8.5 to 9.8 m) above the sea at mean low tide but are occasionally eroded right down to low tide levels and below by battering waves, requiring repairs. This happens even though the rock blocks in the south jetty weigh up to 24 tons (21 metric tons). The jetties were intended to confine and increase the river's current so that it carried sediment a little farther seaward. This has been successful: the Columbia Bar is now 0.5 mile (0.8 km) farther west and in deeper water than in prejetty days. A consequence of the jetties has been the rapid growth of beaches where surf once crashed against bare rock. The North Jetty interrupted the longshore current close to the shore, protecting the shore from erosion by the northerly winter currents and allowing south-flowing summer currents to deposit sediment behind the North Jetty at Benson Beach (stops 2 and 3). Beards Hollow (stop 1), north of the protruding rocky headland of North Head, has also filled in. A long sand beach now connects the cove with Long Beach Peninsula.

Upriver dam construction has brought another unintended change to Cape Disappointment. Humans now regulate the Columbia's discharge and have thus reduced the supply of sediment delivered to the river's mouth. This means less sediment reaches beyond the jetties to be carried northward in winter. Before the jetties were built, some of this sediment would return in summer months with the southbound longshore current. So, beaches such as Benson Beach now receive less material.

Another human alteration is from dredging. The US Army Corps of Engineers removes about 4 million cubic yards (3 million m^3) of sediment per year to improve navigation and deepen the channel between the jetties. Some of this sand is dumped well offshore in deeper water and is lost from the coastal current system, some is disposed in shallow water to be dispersed by waves and currents, and some is disposed in the estuary to create sand islands. As a result, beach growth has slowed. In fact, beaches, especially Benson Beach, have been eroding in recent years.

At stop 1, Beards Hollow Overlook, be sure to read the interpretive sign highlighting some of the changes discussed in this vignette. Then walk to the left, along the cement-block wall to the chain-link fence. As late as the 1930s, much of the swampy, sparsely forested hollow below was a wave-washed cove with a broad, sandy beach. The high wooded knob rising out of the scrub trees and meadow across the hollow is O'Donells Island. In the 1930s it was a sea stack entirely separated from North Head by a channel.

Now head down to Beards Hollow. Continue down the road 0.5 mile (0.8 km) to the parking lot and trailhead at Beards Hollow. The

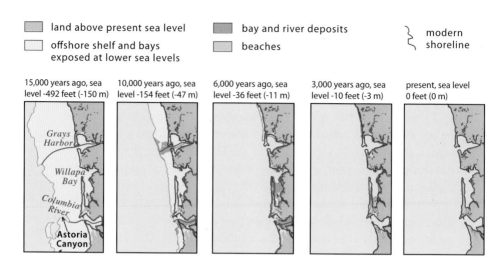

land above present sea level		bay and river deposits		} modern shoreline
offshore shelf and bays exposed at lower sea levels		beaches		

15,000 years ago, sea level -492 feet (-150 m)

10,000 years ago, sea level -154 feet (-47 m)

6,000 years ago, sea level -36 feet (-11 m)

3,000 years ago, sea level -10 feet (-3 m)

present, sea level 0 feet (0 m)

The coastline has migrated eastward as global sea level has risen.
(Redrawn from Twitchell and Cross 2001.)

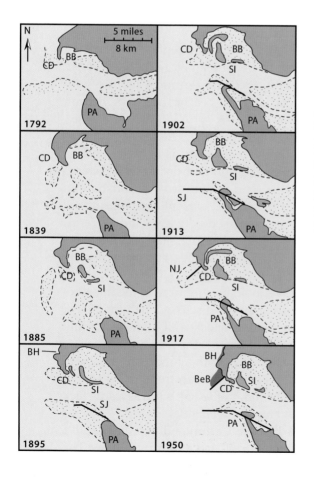

LEGEND

BB	Baker Bay
BeB	Benson Beach
BH	Beards Hollow
CD	Cape Disappointment
NJ	North Jetty
PA	Point Adams
SI	Sand Island
SJ	South Jetty

Changes in the position of shorelines (tan fill) and sandbars (dotted fill) can be traced from nautical charts. Temporary islands are gray, and the accreted sediment at Beards Hollow and Benson Beach are green.

trail in the hollow is flat, paved, and suitable for wheelchairs. About 0.3 mile (0.5 km) from the trailhead O'Donells Island rises on the right side of the path. It consists of pillow breccia—fragmented pillow lava. As sediment swept into this former cove from the north following jetty construction, the beach built outward and the cove became first a salt marsh and then a freshwater marsh just above high tide. As recently as the mid-1950s, O'Donells Island was half surrounded by water at hide tide and people camped on the beach alongside the rock. The shore is now 530 feet (160 m) west of the island. North Head and its lighthouse rise above the south end of this young beach.

Go back out to the highway, turn right, and head to stop 2. Make the 0.25-mile (0.4 km) hike to the top of McKenzie Head for a dramatic contrast with the "handsom view" William Clark and his scouting party enjoyed on November 18, 1805. They first beheld the wild Pacific here, where waves crashed at the base of coastal rocks. Clark wrote: "Those hills are founded on rocks & the waves brake with great fury against them, . . . men appear much Satisfied with their trip beholding with estonishment the high waves dashing against the rocks & this emence ocian."

As late as 1889, the United States Coast and Geodetic Survey's *Coast Pilot of California, Oregon, and Washington* described McKenzie Head as "the first knob to the northwest, three-quarters of a mile from the [Cape Disappointment] Lighthouse. It is an almost round knob, three hundred and fifty yards in extent and one hundred and ninety feet above the sea, covered with grass and fern on top and has no trees. It is almost surrounded by the sea except for a short distance on the northeast side where it is connected with the Cape by a low, sandy neck covered with bushes." If Clark and his band of stalwarts had arrived here today, they would still have to walk another 3,000 feet (900 m) through scrubby vegetation before culminating their journey at the Pacific Ocean. The entire vegetated foreshore of Benson Beach south of North Head has been deposited since North Jetty was completed in 1917.

Now head to stop 3. The pillow basalt in these outcrops is among the best examples you will find anywhere. The rock is resistant basalt of the Eocene-age Crescent Formation, better known from its occurrence in the Olympic Mountains. (Vignette 16 explains the formation of submarine pillow lava and visits more Crescent Formation lava.) The lava in these cliffs erupted underwater off the coast as part of a huge upwelling of magma and was then uplifted above sea level. By late Holocene time these remnants were sea stacks surrounded by surf. Only in the mid-twentieth century, when sand began accreting

Postcard showing Beards Hollow (mistakenly labeled at top) around 1930. O'Donells Island, center, is across the cove. —Courtesy of Lyn Topinka

Beards Hollow from the overlook. O'Donells Island, left of center, is surrounded by the marsh. —Courtesy of John Scurlock

McKenzie Head (extending out of photo at right) protrudes seaward from the low shore in this early nineteenth-century view to the southeast. Cape Disappointment is in the background. Virtually all the beach in this photo now lies well inland. —Courtesy of Washington State Historical Society

Today's view west from the top of McKenzie Head. It is 3,000 feet (900 m) to the shoreline. Waves washed the base of the rocky knob in 1805.

behind the North Jetty, did it become possible to reach the outcrops on foot. They will likely become sea stacks again in the foreseeable future. With the decrease in sediment delivered to this shoreline, and rising global sea levels, these pillows may be inaccessible in a few more decades. There are also fine pillows in the rock wall at the north end of the beach, but high tide might make them difficult to approach.

Stroll along Benson Beach. The sand between the high tide line and the highlands to the east has been piled up about 3 feet (1 m) above sea level by the wind. The head-high bluffs at the top of the sandy beach expose logs trapped in the shallow bay as it was filling with sand to become a shallow lagoon. Subtle layering in bluff cross sections shows the incremental addition of sand. Black magnetic sand on the beach has been eroded from iron-rich Crescent Formation lava.

Erosion is now eating away at the beach and foreshore as much as 15 feet (4.5 m) per year. Enjoy the campground while you can; as of 2012 some of the campsites toward the south end were only 40 feet (12 m) from high tide.

Backtrack out of the campground and head to stop 4, the Lewis and Clark Interpretive Center, which is reached via a paved 500-foot-long (150 m) wheelchair accessible ramped trail. In addition to wonderful interpretive displays about the Lewis and Clark Expedition, there are maps and old photos illustrating historical changes at the coast. The veranda offers ocean views, including McKenzie Head,

Pillow lava in the former sea stacks at stop 3.

to the north. Consider walking 100 yards (90 m) north on the Cape Disappointment Trail for even better views to the north; this trail leaves the paved ramp on the north side of the visitor center.

Cape Disappointment Lighthouse, constructed in 1856, and the rocky prominence it rests on block the southern view. Two trails lead to the lighthouse across the narrow Dead Mans Cove. From the lighthouse you have a view of the entire Columbia River mouth and far upstream, as well as south down the Oregon coast. Both jetties stretch far out to sea: North Jetty is 2.4 miles long (3.9 km) and South Jetty is 6.6 miles (10.6 km) long. Nearly constant shipping traffic demonstrates the importance of the river to commerce and the need for safe passage across the bar. The geomorphic changes we have seen today, including this heavily traveled river mouth, are the consequence of human geologic engineering. But hey, who doesn't like sandy beaches?

Shoreline changes from 1869 to 2012 taken from maps and aerial photographs. In 1951, 4,200 feet (1,300 m) of land had been added west of McKenzie Head since Clark's visit in 1805. In 2012, the distance was 2,600 feet (800 m). Between 1999 and 2012, the shoreline near the campground eroded 145 feet (44 m). (Adapted from Kaminsky et al. 2010.)

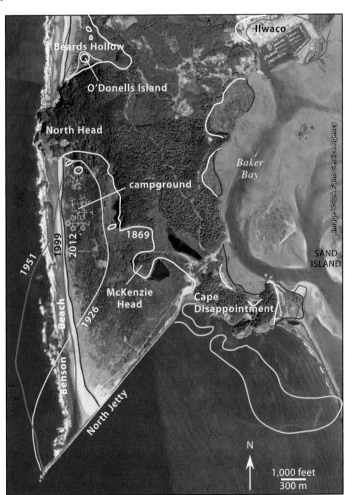

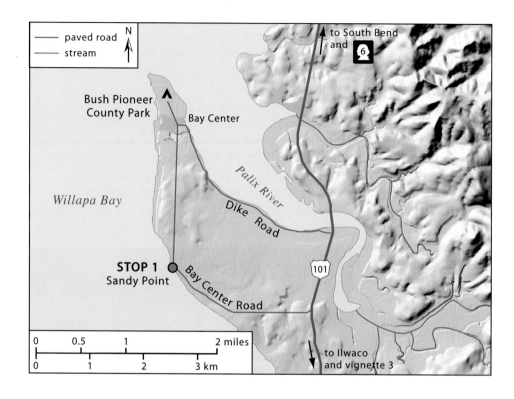

GETTING THERE: Sandy Point, the only stop in this vignette, is located along Willapa Bay in southwestern Washington. From Aberdeen, follow US 101 south for 42 miles (68 km). One mile (1.6 km) south of the bridge over the Palix River, turn west (right) onto Bay Center Road (not Bay Center Dike Road!). You reach the shore of Willapa Bay after 1.7 miles (2.7 km), at a small paved pullout on the west side of the road. Park here and walk the shoulder of the lightly traveled road for 100 yards (100 m), or drive to the north end of the guardrail where there is a narrow shoulder with space for one car. This is Sandy Point. Consult tide tables, available online, for Bay Center in Willapa Bay to determine the best times to visit Sandy Point. The beach and its stumps are best exposed with a 4-foot or lower tide; visit during an ebbing tide to have the most time to explore the site.

4

Witness to a Giant Earthquake

THE SUNKEN FOREST OF WILLAPA BAY

Receding tides at Sandy Point and other locations around Willapa Bay reveal ragged weather- and tide-beaten tree stumps, the roots of which spread across intertidal sand and tide pools. Rockweed and other algae drape the stumps, and barnacles and oysters cling to some. At the north end of the guardrail, scramble down the low wall of basalt riprap to the sand beach. The beach is firm, but you can expect to get wet feet from puddles. As you walk across the tidal flat at Sandy Point, note that many stumps are small, less than 1 foot (0.3 m) across, but there are much larger ones up to 3 feet (1 m) across. The latter stumps are remnants of once-mighty spruce trees. What is going on here? Why are tree stumps below the high tide line?

The spruce stumps at Sandy Point are probably among the tens of thousands of arboreal victims of a giant earthquake that struck the coast of the Pacific Northwest in January 1700. During this earthquake, the leading edge of the North American Plate, from Vancouver Island to Cape Mendocino, lurched seaward an esti-mated 60 feet (18 m) along the Cascadia Fault. One of the effects of this sudden shift was the down-warping, or subsidence, of much of the Pacific coast in this region. The land fell about 5 feet (1.5 m), enough to drop the lowest parts of coastal forests into tidewater. The trees, unable to tolerate the salt, soon died. A coastal marsh formed, and fresh sediment covered the stumps. Coastal erosion during the past few decades has exhumed the stumps; evidence of this erosion is plain to see in the dying trees along the road at the head of the beach, which have contributed a pile of recently fallen logs to the shore.

Almost three centuries after the quake, scientists used such trees to date the earthquake. Though they did not use the spruce stumps at Sandy Point, radiocarbon ages obtained from others nearby show that the trees died between 1695 and 1710.

Scientists further narrowed the earthquake's time window with help from western red cedars. The wood of this species is well-known

Stumps of spruce trees on the beach at Sandy Point.

for its resistance to decay; it is prized for use as fence posts, shingles, and decking. Resistance to decay has allowed trunks of western red cedar, though killed in 1700, to remain standing today in tidal marshes of estuaries in southwest Washington. One such grove of snags is only 3 miles (5 km) from Sandy Point along the South Fork Palix River. The growth rings in these snags were compared with those of old-growth trees cut on Long Island, in Willapa Bay, in 1986 and 1987. The comparison showed that the dead red cedars died after the end of the growing season of 1699 but before new wood was added in the spring of 1700.

The stumps at Sandy Point tell part of a detective story. Most of the clues in this story were unearthed along the Pacific coast of North America, where the earthquake spawned a tsunami that

A dark layer of organic soil, formerly the forest floor, survives wave erosion and lies beneath the stumps.

washed over Native American fishing camps, including one along the nearby Niawiakum River 2 miles (3 km) northeast of Sandy Point, and drowned thousands of shoreline trees. But pivotal clues also come from Japan.

Written records from Honshu, Japan's main island, tell of a tsunami that flooded villages and fields at six places scattered along more than 500 miles (800 km) of the island's east coast. The flooding began around midnight (Japan time) on January 28, 1700, and lasted for twelve hours or more. While only two people are known to have died from the tsunami, crops were lost to saltwater flooding, houses and salt kilns were destroyed, and many people were left homeless. The written records do not give the heights reached by the tsunami, but the effects described suggest 5 to 15 feet (1.5 to 4.5 m).

The headman of Miho, a village southwest of Tokyo, wrote of a series of seven sea surges well above normal high tide levels that receded with powerful currents he likened to rivers. The origin of the waves puzzled him. He knew that tsunamis were usually associated with earthquakes, but he had felt no shaking.

The "orphan tsunami" of 1700 was thus a mystery for the residents of the Japanese coast who suffered from it, and for later Japanese historians and earthquake researchers. It wasn't until the mid-1990s that US Geological Survey researchers, using evidence such as the drowned forests of Washington's coast and the orphan tsunami, solved the mystery. Allowing for modern timekeeping methods and time zones, they estimated that a huge earthquake off the west coast of North America occurred at 9 p.m. on January 26, 1700, spawning the January 28 tsunami, which reached Japan ten hours after the quake.

But why did the trees in Sandy Point drown in the first place? To understand why, we need to explore the tectonics of the Cascadia Subduction Zone. The westward-moving North American Plate overrides the Juan de Fuca Plate below the surface along the coast at a rate of about 1.2 to 1.6 inches (3 to 4 cm) per year. The two plates generate friction as they grind past each other and become locked. In effect, the relatively thin leading edge of the North American Plate is imperceptibly dragged inland by this convergence friction. How much is it dragged per year? Using high-precision GPS methods, scientists estimate the reverse movement to be as much as 1 inch (2.5 cm) per year. But most of the North American Plate is a huge mass of thick continental crust extending east all the way to the middle of the Atlantic. The mass of this plate can't be easily moved. So, in response to this convergence friction, the crust deforms vertically, or is uplifted, 0.03 inch (0.8 mm) or more each year. This may not sound like much of a change in surface elevation, but in the timescales that geologists think in, this is a very fast uplift rate, sufficient enough to cause high mountains to grow. (See vignette 15 to explore evidence of uplift on the coast.)

There are no high mountains here along the coast (the Olympics are a different story; see vignette 16) because every now and then the friction is overcome and the locked plates release, generating a huge earthquake; the vertical deformation relaxes, bringing the uplifted coast back to, or even below, sea level. The cycle renews when the plates again lock up in their titanic, slow collision.

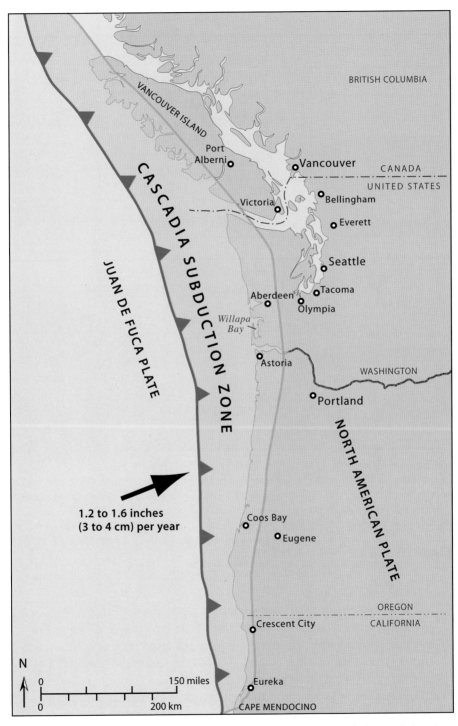

BRITISH COLUMBIA

VANCOUVER ISLAND

Port Alberni

Vancouver

CANADA
UNITED STATES

Victoria

Bellingham

Everett

CASCADIA SUBDUCTION ZONE

JUAN DE FUCA PLATE

Seattle

Tacoma

Aberdeen

Olympia

Willapa Bay

Astoria

WASHINGTON

Portland

NORTH AMERICAN PLATE

1.2 to 1.6 inches
(3 to 4 cm) per year

Coos Bay

Eugene

OREGON
CALIFORNIA

Crescent City

N

0   150 miles

0   200 km

Eureka

CAPE MENDOCINO

The Juan de Fuca Plate, beneath the Pacific Ocean, moves eastward beneath the North American Plate; the red, toothed line indicates the plate boundary off the coast. Friction between the plates creates a locked zone (orange area) along the entire contact surface of the two plates.

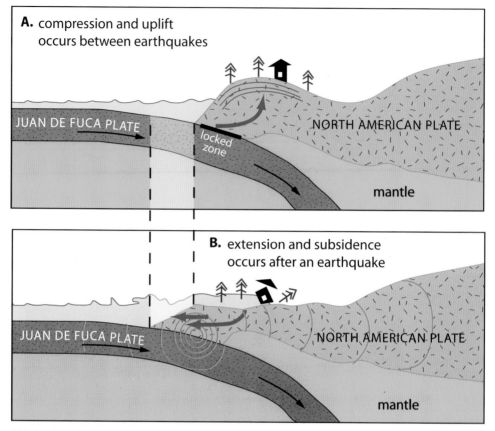

A. compression and uplift occurs between earthquakes

JUAN DE FUCA PLATE

locked zone

NORTH AMERICAN PLATE

mantle

B. extension and subsidence occurs after an earthquake

JUAN DE FUCA PLATE

NORTH AMERICAN PLATE

mantle

(A) As the eastward-moving Juan de Fuca Plate is subducted beneath the westward-moving North American Plate, friction creates a locked zone between the two plates. The massive continental plate, resistant to reverse motion, deforms upward to accommodate the stress. (B) Eventually the locked zone ruptures and generates an earthquake, partially or entirely relaxing the uplifted area and allowing the leading edge of the continental plate to rebound westward. The vertical dashed lines represent this rebound. In general, the greater the area that is unlocked, the greater the earthquake. (Modified from Hyndman et al. 1996.)

Now we see how the land surface can be raised a little above sea level, allowing marshes or even forests to grow, only to suddenly subside, even below sea level, drowning the forests. How often do these great earthquakes occur? At several sites near Willapa Bay and Grays Harbor, riverbanks expose repeated geologic sequences of intertidal mud separated by thin layers of buried soil with plants and tree stumps. Each of these soil layers bear witness to a cycle of uplift and sudden submergence. North of Willapa Bay, along the banks of the Johns River at Grays Harbor, seven buried soils, each radiocarbon dated, span the period from 1400 BCE to 1700 CE. The

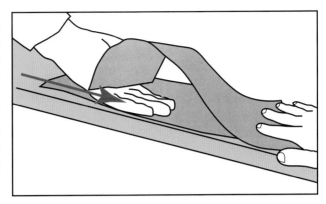

Slide your hand at a downward angle under the edge of a sheet of paper; the friction causes the paper to bend upward. This is analogous to the uplift that occurs in the thin leading edge of the North American Plate (the paper) as the Juan de Fuca Plate (your hand) is subducted below it.

six time intervals between the soils average close to five hundred years. But the intervals are not regular; the time between inferred earthquakes ranged from a few centuries to a full millennium. All of these earthquakes were probably of magnitude 8 or larger, and the 1700 Cascadia earthquake was almost certainly magnitude 9.

Just how violent was the 1700 Cascadia earthquake? Earthquakes are usually measured in terms of magnitude, or the amount of energy released at the time of fault rupture. Seismologists use the moment magnitude scale to measure the absolute amount of energy released during an earthquake. Magnitude is expressed in whole numbers and decimal fractions in a logarithmic scale. A moderate earthquake may have a magnitude of 4.4, while a stronger earthquake might be rated a 5.4. The magnitude increase with each whole number is ten times the magnitude of the preceding whole number, with an energy release about thirty-one times more than the preceding whole number value.

You might be familiar with other measures, such as the Richter and Mercalli intensity scales. The Richter scale, devised in the 1930s, is the ancestor to the moment magnitude scale. It also measures the energy released but doesn't adequately measure the magnitude of the largest earthquakes. Media reports of earthquakes often cite a Richter magnitude even though earthquake observatories calculate and report the moment magnitude almost exclusively. The older Mercalli scale subjectively measures the severity of shaking at a given location and its effect on buildings, people, and the environment based on individual testimonies.

One clue to the size of the Cascadia earthquake comes from the orphan tsunami: the wave heights recorded in Japan that January evening. Computer models have been used to estimate the

900 CE
inconspicuous here

1700 CE

700 CE

400 CE

400 BCE

900 BCE

tidal river

Exposed soils (dashed lines) in a bank of the Johns River record periods in which the ground was up-lifted above sea level, followed by sudden subsidence to intertidal, or lower, elevations. The layers be-tween the soils are intertidal deposits. Dates refer to calendar years. An additional soil dated to 900 CE is inconspicuous in this photo. An older seventh soil was below the tide level at the time the photo was taken. A hoe with striped handle provides scale, far left. —Courtesy of Brian Atwater, US Geological Survey

magnitudes of the earthquakes that spawn such ocean-crossing waves. Couple the Japanese tsunami heights with the measured displacement that occurred along the Cascadia Fault in 1700, add in the length of the fault surface that likely ruptured, and a moment magnitude (M) can be estimated: between M 8.7 and M 9.2.

It was a monstrously big earthquake that ranks among the largest ever recorded. Only seven earthquakes in all recorded history are calculated to be M 9 or larger, including the January 1700 Cascadia earthquake and the Sumatra earthquake of 2004, which caused the huge tsunami that killed more than two hundred thousand people the day after Christmas. The amount of energy released in an M 9 earthquake is estimated to be about 4 trillion kilowatt hours. For comparison, in one month the United States uses 2.5 trillion kilowatt hours of energy.

SIZE RANKING	DATE	LOCATION	MOMENT MAGNITUDE
1	May 22, 1960	Valdivia, Chile	9.5
2	March 27, 1964	Prince William Sound, Alaska	9.2
3	December 26, 2004	Sumatra, Indonesia	9.1
4	March 11, 2011	Northeast Japan	9.0
5	November 11, 1952	Kamchatka, Russia	9.0
6	August 13, 1868	Arica, Peru (now in Chile)	9.0
7	January 26, 1700	Cascadia Subduction Zone	9.0
8	February 27, 2010	Maule, Chile	8.8
9	January 1, 1931	Esmeraldas, Ecuador	8.8
10	February 4, 1965	Rat Islands, Alaska	8.7

(List from the US Geological Survey Earthquake Hazards Program.)

The ten largest earthquakes in recorded history.

What was the effect of this giant earthquake on the northwest coast? It warped the seafloor upward as much as 20 feet (6 m) all along the boundary between the North American and Juan de Fuca Plates. Many areas behind this plate boundary subsided suddenly 5 feet (1.5 m) or more, sometimes to below sea level, as did the former forest floor we see at Sandy Point. Destructive shaking probably continued for several minutes; people living along the coast were likely unable to remain standing. The displacement of the seafloor generated a tsunami that probably reached heights of 50 feet (15 m) along some parts of the Pacific coast between Vancouver Island and Cape Mendocino. Computer simulations show that the flooding in that region began twenty minutes after the earthquake.

While part of the 1700 tsunami was coming ashore in the Pacific Northwest, another part of it was racing across the Pacific Ocean. Out there, at water depths of 2.5 miles (4 km), a tsunami travels at jetliner speeds. It would not have been noticeable in the middle of the ocean because the crest-to-crest distance (the wavelength) was probably 50 miles (80 km) or more. But when these long-period waves approached shallow water off the coast of Japan, they slowed down and built up enough to enter Japanese written history.

Tsunami waves are normally not the high wall of breaking water typically portrayed in warning signs and the media. They may not appear as a wave at all, but rather as a series of sudden surges sweeping inland, reaching well above normal tide levels and swamping low-lying inland areas. Each surge is followed by the next, and then the next, each pushing farther inland on the back of the preceding one. The water recedes with powerful currents, pulling houses, debris, and people with it. During the dark night of January 26, 1700, the sequence of sea surges washed inland along the Cascadia coast, carrying sand and marine life with them, drowning freshwater marshes with salt water. The salt water killed plants and damaged soils. There were surely many drowned animals and people.

Stories of waves, subsiding land, and earthquakes are found in the oral traditions of native people up and down the Pacific coast, from the Yuroks of northern California to the Kwakwaka'wakw (Kwakiutl) of northern Vancouver Island, an area that approximates

These signs are proliferating along the Pacific coast. Tsunamis are generally rapid surges sweeping inshore rather than giant waves, as depicted here. The necessary response is the same.

the extent of the Cascadia Subduction Zone. Oral histories of the Hoh, Quileute, and Makah along the Washington coast may relate to earthquakes and tsunamis. Though these cannot be dated with certainty, at least some are likely to refer to the giant earthquake and tsunami of 1700. Both the Hoh and Quileute tribes of the Olympic Peninsula tell of a great struggle between the mythical Thunderbird and Whale. One account states: "The great bird grasps the giant orca in his talons and the resulting battle causes a great shaking, jumping up and trembling of the earth beneath, and a rolling up of the great waters." A Makah story tells of the sea rising and flooding across the low estuary between the outer Pacific shore and Neah Bay, turning Cape Flattery into an island.

Oral histories collected from Yurok elders of the northern California coast told of two mythical characters, Earthquake and Thunder, who made the ground shake, sink, and then submerge beneath the ocean. In these stories researchers found references to specific locales where they found tsunami deposits from the early 1700s.

This illustration by Edward Malin reproduces a late nineteenth-century ceremonial painting in a Nuu-chah-nulth longhouse on Vancouver Island. The depiction of Thunderbird carrying Whale is a common northwest native allegory for earthquakes and tsunamis. —Courtesy of Edward Malin

We've seen enough recent M 9 earthquakes to have a good idea of what the next one will do to Cascadia. The Cascadia Region Earthquake Workgroup (CREW) is a coalition working to reduce the effects earthquakes have on Cascadia region communities. CREW prepared a booklet, including photos from historic quakes, to bring the dreadful scenario of an M 9 earthquake to life. Among their conclusions:

- The earthquake will affect every business, government agency, nonprofit organization, and individual in the region.

- Strong shaking will last for four minutes, possibly killing and injuring thousands.

- A tsunami up to 33 feet (10 m) high will wash over the coast within minutes.

- Much of coastal US 101 will be impassable due to wave and landslide damage.

- Parts of the coast will be cut off from inland cities. Roads through the Cascades may likewise be blocked.

- Most places will be on their own for rescue and medical help.

- Utilities and transportation in the I-5/Washington 99 corridor will be disrupted for months.

- Cities may have significant fatalities as tall buildings collapse.

- Aftershocks will continue for years; some of these could be large earthquakes.

Many people living along the inland shoreline think that the distance between them and the open Pacific coast will protect their coastal towns and beach homes from a tsunami generated by a Cascadia Subduction Zone earthquake; however, the Strait of Juan de Fuca and the inland waters of the Salish Sea might not be protected. A train of tsunami waves that reaches clear across the Pacific to Japan would also rush up the Strait of Juan de Fuca. Computer simulations clearly show that waves could reach places like Port Angeles ninety minutes after an earthquake. A layer of marine sand found in several freshwater marshes on northern Whidbey Island is probably geologic evidence for an earthquake-generated tsunami.

On a broad scale, waves reaching inland would likely be attenuated; however, modeling suggests that the confined inland waters may concentrate the waves in some places, such as the heads of shallow bays, and cause them to grow even higher. Models show the first tsunami crest striking the Anacortes-Bellingham area within about two hours of an M 9 earthquake. Waves as high as 11 feet (3 m) could swamp the Nooksack River delta in Bellingham Bay and inundate low-lying coastal communities, such as Edison on the

Samish River delta, Marietta near Bellingham, and even Ferndale (via the Nooksack River). The tsunami would continue southward into Puget Sound and northward toward the lower mainland of British Columbia, although simulation models are incomplete for inundation in those areas.

Fortunately, there should be a warning of an approaching tsunami, robably in the unfortunate form of the huge earthquake itself, and if infrastructures survive the shaking, by follow-up announcements and warnings. The National Tsunami Warning Center is responsible for issuing warnings about potential tsunamis along the west coast of the United States. They issue bulletins within fifteen minutes of large undersea earthquakes in the Pacific Ocean basin. In the meantime, public awareness and education is of great importance. If you live in the hazard zone of a great Cascadia Subduction Zone earthquake, are you prepared?

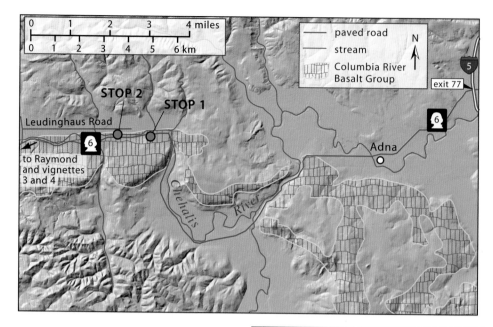

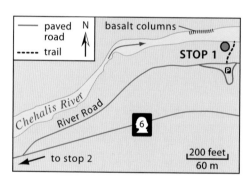

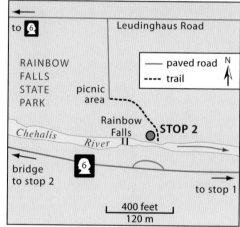

GETTING THERE: Take exit 77 from I-5 (at Chehalis) and head west on Washington 6. After 14 miles (22 km), turn right on River Road and proceed 0.1 mile (0.2 km) to a parking area on the right; this is the former site of a bridge over the Chehalis River. A path goes about 3 feet (1 m) down the bluff to the river. If you mistakenly turn at the first River Road junction 12 miles (20 km) from I-5, you'll reach stop 1 after 2 miles (3 km) or so, only from the east.

To reach stop 2, return to Washington 6 and continue west 3.2 miles (5.1 km), then turn right onto Chandler Road. Cross the river and turn right on Leudinghaus Road. Proceed 0.8 mile (1.3 km) to the park entrance. Stop at the picnic area and walk to the falls. Find a spot to drop over the bank to the rocks on the shore. An alternative stop 2 is on the south side of the river, but it is more difficult to examine the basalt in any detail there. To get there, starting from stop 1 travel west on Washington 6 for 2.3 miles (3.7 km) and stop at the signed pullout for Rainbow Falls. Walk the short path from the road to the rock exposed beside the river.

5

The Columbia River Basalt
in the Chehalis River

The stops in this vignette visit remnants of basaltic lava that flowed here all the way from the Grande Ronde country in northeast Oregon, a distance of around 350 miles (560 km). The lava is what remains of one of the three known McCoy Canyon basalt flows, which are part of a larger package of lavas called the Sentinel Bluffs Member of the Grande Ronde Basalt Formation, which are in turn a subgroup of the Columbia River Basalt Group. This group comprises as many as three hundred individual lava flows that combined represent one of the largest outpourings of lava on Earth.

The McCoy Canyon lavas erupted 15.5 million years ago. These enormous basalt flows cover 33,000 square miles (82,460 km²) with a volume of some 1,000 cubic miles (4,280 km³). The individual flows are 30 to 400 feet (10 to 120 m) thick in the Pasco basin, near the Tri-Cities in south-central Washington. The base is not exposed near Rainbow Falls, but the flow is probably less than 100 feet (30 m) thick here, near its leading edge.

The Columbia River Basalt Group is a tremendous stack of lava flows derived from truly enormous outpourings of basalt that buried much of eastern Washington and parts of Oregon and Idaho. Thick deposits of basalt derived from such long-term, large-scale eruptions are called flood basalts. The Columbia River Basalt Group flood basalts began erupting in the Miocene epoch, about 17 million years ago, and continued intermittently over the next 11 million years; about 96 percent of the lava erupted in the 2.5-million-year interval between 17 and 14.5 million years ago. In most cases, eruptions began with fountains and sheets of lava erupting from fractures, known as fissures, which were 50 to 60 miles (80 to 100 km) long. In the latter stages of individual eruptions, as the rate of lava effusion declined, parts of these fissures became plugged, and eruptions became concentrated at discrete vents. Later, younger lava flows often buried these volcanic structures, so few are exposed for

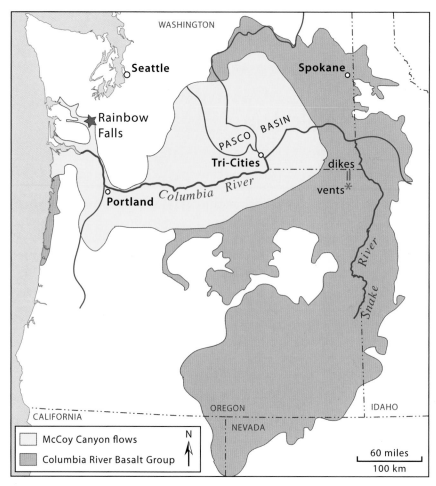

The area covered by lava flows of the Columbia River Basalt Group is shown in brown. Chemical analyses of the McCoy Canyon flows (yellow) show that they erupted from dikes and vents in northeast Oregon. Erosion or later flows have covered the connection between the flows and the eruption areas. (Adapted from Reidel 2005.)

study in the source area of this great flood basalt. Much smaller fissure eruptions occur today at volcanoes such as Kilauea in Hawaii and Réunion in the Indian Ocean.

The precise reasons for this great lava effusion are hotly argued. Some chemical and geophysical evidence suggests that it may be related to magma from the Yellowstone hot spot. This plume of hot, partially molten mantle rock is rising into the crust from perhaps 300 miles (480 km) below the surface. The hot spot is currently under the caldera at Yellowstone National Park, but the North American Plate is drifting west-southwest over the plume. Around 16.5 million years ago, it was below southern Oregon, and its blowtorch effect may have caused the eruptions of the Columbia River Basalt Group. The surface manifestation of the hot spot will continue migrating northeast across Montana, reaching into central Canada in many millions of years.

The eruptions ponded in and filled the Pasco basin with as much as 2.5 miles (4 km) of basalt. Most of these lava flows found their way out of the basin by way of the ancestral Columbia River valley, which was much wider then. Some made it all the way to the Pacific coast; a few found their way northward to Grays Harbor in long-gone river valleys. The lava in the Chehalis River valley is the most easily accessible of these northwestern lavas flows, but scraps are found another 30 miles (48 km) to the northwest.

Where topography was subdued, the lava flows could spread out and cover the landscape in sequential sheet flows. One flood of lava would cover a large area and cool to form a relatively high lobe of rock; the next great lava flood would be channeled alongside this new topographic high and at least partially cover most of the older flow. Subsequent flows lay alongside or overlapped each other like an untidy pile of pancakes. If the lava flowed into a river valley, however, the valley walls channeled it, and the lava cooled as a much thicker mass since it couldn't spread out. These are called intracanyon lava flows, of which the lava in the Chehalis River valley is one.

Stop 1 is at the former site of a one-lane steel bridge. Upstream you can see 6-foot-high (1.8 m), 2- to 3-foot-wide (0.6 to 0.9 m) lava columns lining the opposite bank of the river. Scramble down the bank to the water's edge. At low water, in summer and fall, it is easy to walk along the river's edge to see the columns better. The water-smoothed rock along the shore consists of eroded cross sections of the crudely shaped basalt columns.

The waterworn surface of the riverside basalt is gray. Break open a sample to expose a fresh surface. The fresh rock is very fine grained and dark gray to black. In decent light you may see sparkles from tiny plagioclase crystals, which are needle-shaped or blockier prisms. Even the largest crystals are less than 0.04 inch (1 mm) long. The rest of the rock, or the groundmass, is amorphous even under a hand lens. It is composed of microcrystals, visible only under a microscope, and glass, the chilled, liquid portion of the lava that failed to form crystals. Vesicles, the air pockets left behind by gas trapped in the lava, may be evident with a hand lens, though some are conspicuously larger on weathered surfaces. Some vesicles are semispherical, but others have irregular shapes where they were

At stop 1, a row of lava columns, each about 3 feet (1 m) across, is exposed in the bank of the Chehalis River upstream. They are mostly covered with brown moss above the high water level.

confined between growing microcrystals. Many loose fragments of the basalt may have a brown rind nearly 0.4 inch (1 cm) thick, the result of the past 16 million years or so of weathering.

The shape of the columns across the river is a clue to how deeply the lava has been eroded at this location. The classic Columbia River Basalt Group lava flow is typically divided into more or less horizontal zones, each characterized by a distinct pattern of fractures. These result from varying cooling styles and cooling rates within the lava. The base of a flow often has stout columns bound by vertical fractures that formed as cooling fronts migrated upward into the lava from the much colder ground the lava flowed over. Colder solids take up less space than hot ones, so the space given up through contraction was taken up by the fractures (see vignette 22 for more on columns like this). The basal portion of the flow is called the colonnade. These columns tend to be wide, up to 3 feet (1 m), due to the slow migration of the cooling front into the flow once it had stopped flowing. The uppermost part of the lava flow, the flow top,

consists of either rubbly breccia or a relatively smooth, solid surface with ropelike flow structures. The style depends on a variety
of factors, chief among them velocity and viscosity. Regardless of
style, the flow top is typically rich in vesicles, which formed from
gas bubbles trapped in the crystallized lava, and may have crude
colonnade-style columns.

The zone between the lower colonnade and the flow top is usually the thickest of the three. Called the entablature, it is characterized by chaotically intersecting fractures and dense rock that has far
fewer vesicles than the flow top. The interior of these flows cooled
much more slowly, perhaps taking months or even several years if
the flow was especially thick. The entablature jointing style is typical of Columbia River Basalt Group lavas, though it is quite uncommon in lavas elsewhere. The reason for this, or for the chaotic nature
of entablature jointing itself, is the subject of much speculation and

*Cross sections of columns under the former bridge have several sides. The classic shape is hexago-
nal, but the outlined column appears to have seven, which is not unusual.*

conjecture, and it has yet to be satisfactorily resolved. The intersection of many fractures is probably related to the very slow progression of cooling into the lava from both the top and bottom, and perhaps the sides.

These distinct cooling zones can easily be recognized in the nearly endless stacks of lava flows lining the highways in dry, relatively vegetation-free eastern Washington. Individual flows are not as easily traced here in the heavily forested Chehalis River valley, where erosion is also far more intense. The columns at stop 1 are a remnant of the lower colonnade of one of the McCoy Canyon flows. The flow is found on the ridge crest south of the Chehalis River, at around 1,050 feet (320 m) of elevation, or nearly 800 feet (245 m) above the falls at stop 2. However, this does not represent the actual thickness of the basalt here, since the rock layers of the ridge have been tilted up to the south. The remaining basalt is probably no more than 70 to 80 feet (21 to 24 m) thick.

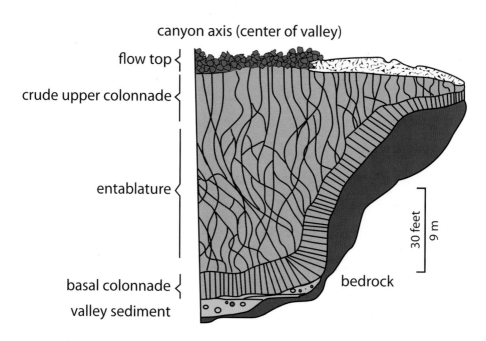

An idealized intracanyon Columbia River Basalt Group lava flow. A thick, rubbly flow top is on the left, and a smoother surface on the right. A single flow could have both surface types. Joint patterns define different cooling regimes within a flow. The columns at stop 1 are like those depicted in the base of the figure, and the blocky fracture pattern seen at Rainbow Falls (stop 2) is from the entablature, a little higher in the flow. (Adapted from Tolan et al. 2009.)

Stop 2 is at Rainbow Falls. (Remember to leave your rock hammers in the car when at this or any state park.) The Chehalis River passes over a low black rock step of the same lava flow exposed downstream at stop 1. A myriad of intersecting fractures, characteristic of the entablature of the Columbia River Basalt Group, breaks the basalt into small chunks. The flow top, and an unknown amount of the flow's interior, has been stripped away by erosion. There is no evidence that glaciers reached this far south during the Pleistocene, so flowing water was the agent that planed the lava flow to its current level.

Chunky blocks of basalt at Rainbow Falls State Park. Pocket knife for scale.

Rainbow Falls tumbles over a resistant ledge of basalt.

Thick sediment overlies the basalt along the riverbank; it is especially noticeable on the north bank. The existence of the falls tells us
that erosion is not finished here; the river is still trying to cut its way
down to the elevation of its mouth at Grays Harbor, in Aberdeen. By
the time it has done that, perhaps there will no longer be a remnant
of the McCoy Canyon flow in the river channel.

How long did it take for this lava flow to reach this place, so far
from its source? Basaltic lavas advance at various speeds. Velocity is
controlled by a number of factors. Foremost among these is the viscosity of the magma as it is erupted. Viscosity is defined as the ability of a substance to resist flow. A substance with higher viscosity is
less likely to flow easily. For example, honey is more viscous than
water; thus it flows more slowly. Viscosity is itself dependent on the
strength of the chemical bonds within a substance. Other important
factors that affect the speed of lava are the slope of the land it flows
on; whether or not it encounters water, which cools it; the distance
it travels; and the amount of insulation the cooler lava-flow surface

FACTOR	EFFECT ON VISCOSITY
Increase in temperature	Decrease: Molecular bonds weaken at higher temperatures.
Increase in silica content	Increase: Chains of silica molecules thicken lava.
Increase in water content	Decrease: Water decreases the strength of molecular bonds.
Increase in crystal content	Increase: Solids thicken lava.
Increase in gas content	Decrease (mostly): Dissolved gases interfere with molecular-bond formation. However, once lava begins to get foamy, viscosity increases—as do eggs when whipped with a mixer.

Variables that affect lava viscosity. An increase in viscosity means that lava is pastier and flows less readily.

provides the hot interior. Viscosity increases as lava cools during flow, to the point that it no longer flows.

The rate of emplacement of Columbia River Basalt Group lavas, or the time it took them to flow the greatest distance from their various vents, is a controversial subject. Over the years scientists have postulated a number of scenarios, based partly on historical observations of much smaller lava flows and partly on mathematical models that describe lava behavior based on measurable factors, such as crystal and silica content, and estimated factors, such as water and gas content. Eruptive temperature is very important in these scenarios, but that can be estimated from temperatures of modern lavas with similar chemical compositions.

An early study, in 1970, proposed that Columbia River Basalt Group flows were of very low viscosity and perhaps 15 to 300 feet (5 to 100 m) thick. They moved as quickly as 0.5 to 1.25 miles (1 to 2 km) per hour and took only days to a few weeks to cover hundreds of kilometers. That is extremely fast for a lava flow. The fastest low-viscosity Hawaiian lava yet recorded erupted from Mauna Loa in 1950. It moved at a similar rapid clip—5.8 miles (9.3 km) per hour—but only for a short period of time and down steep slopes. Most Hawaiian flows are far slower, on the order of 0.6 mile (1 km) per hour. At the more typical rate, it would take a Hawaiian basalt lava twenty-four days to travel the 350 mostly flat miles (560 km) covered by the McCoy Canyon lava. Of course, the huge volumes of lava that poured out of fissures (now dikes) during the Columbia River Basalt Group eruptions were a major driving force behind their spreading rate; nothing like those volumes erupt in Hawaii, or anywhere else.

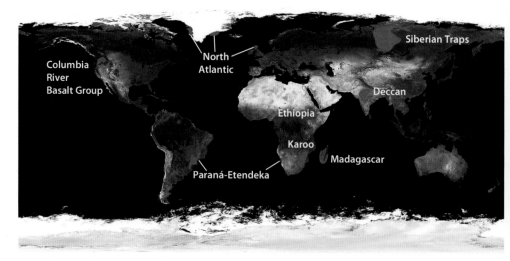

Present-day exposures of the major continental flood basalts of the past 250 million years. The Columbia River Basalt Group appears large compared to others with greater volumes because it is the youngest and least eroded. Its original extent is still relatively well defined.

A 1996 model proposed flow rates of around 0.2 mile (0.4 km) per hour. At that rate, it would have taken the McCoy Canyon lava seventy-three days to reach Rainbow Falls. The most recent studies, based largely on the chemical composition of the lavas and their crystal content, have come full circle, suggesting rapid emplacement over days to perhaps a month for the Sentinel Bluffs Member, including the McCoy Canyon lava. Even at a "fast" flow rate of 0.6 mile (1 km) per hour, you would be able to outwalk one of these lavas, and you would sure want to!

The Columbia River Basalt Group lava flows, unbelievably huge as they may seem, are dwarfed by a number of much larger flood basalts. The largest known continental flood basalt is the Siberian Traps (*trap* is an older term for basalt and related dark, fine-grained igneous rocks). These lavas erupted about 250 million years ago and had an estimated volume of around 700,000 cubic miles (3 million km³), enough to bury the states of Washington, Oregon, Idaho, California, Nevada, Utah, and Arizona beneath 1.3 miles (2 km) of lava, or the entire land surface of the planet with nearly 900 feet (270 m) of lava!

The barely conceivable volume of lava in the Siberian Traps is responsible for the greatest mass extinction suffered by life on Earth, which happened at the end of the Permian period roughly 252 million years ago. About 95 percent of all marine species died out in a geologic instant of perhaps 60,000 years. Terrestrial vertebrates, insects, and plants suffered very large extinctions. The killing mechanism is not fully understood. A widely held scenario involves the volcanic winter hypothesis. This idea maintains that great masses of sulfur dioxide and aerosolized compounds enter the atmosphere, released from periodic eruptions of really huge volumes of lava. These compounds increase the planet's albedo, meaning that more sunlight is reflected back into space. Consequently, photosynthesis is severely reduced or stopped, and the lower stratosphere is cooled, too.

An alternative hypothesis suggests that global warming, rather than cooling, would have resulted from the release of a great volume of the greenhouse gas carbon dioxide, which traps heat in the atmosphere. Carbon dioxide is a by-product of volcanic eruptions, and it doesn't break down easily, unlike the aerosols blamed for volcanic winters. Since plants convert carbon dioxide to carbon and oxygen, a die-off of plants due to the earlier, faster accumulation of global-cooling gases would allow carbon dioxide to accumulate in the atmosphere. A synthesis of these apparently competing atmospheric effects produces an alternating cycle of cold, dark winters and long, hot summers brought about by the on-again, off-again eruption of large volumes of basalt. Suffice it to say that scientists don't yet understand the precise relationship between mass extinction and the eruption of flood basalts. The Columbia River Basalt Group is not associated with a planetary mass extinction event, probably because of its relative small size.

Province	Age*	Present volume (miles³/km³)	Original volume (miles³/km³)	Duration**
Columbia River	15	40,000/180,000	40,000/180,000	about 1 (for 90 percent)
Deccan, India	66	125,000/520,000	240,000 to 480,000/ 1 to 2 million	about 1
Paraná-Etendeka, South America/Africa	132	20,000/80,000	360,000/1.5 million	1 to 5
Karoo, South Africa	183	not known	240,000 to 480,000/ 1 to 2 million	0.5 to 1
Siberian Traps	250	80,000/340,000	700,000/3 million	about 1

*millions of years ago   **millions of years

Major continental flood-basalt provinces of the past 250 million years.

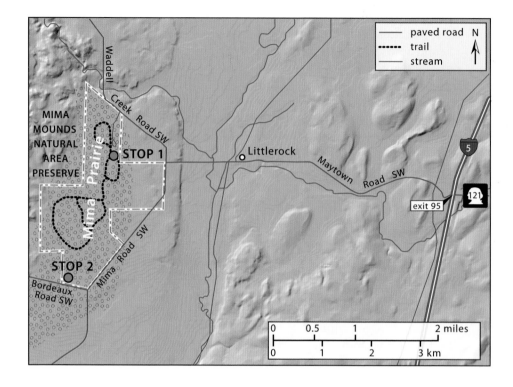

GETTING THERE: The Mima Mounds Natural Area Preserve is reached from exit 95 on I-5. Head west on Maytown Road SW through the town of Littlerock. To reach stop 1, after 3.8 miles (6.1 km) turn right at the T intersection with Waddell Creek Road SW. After 0.8 mile (1.3 km) turn left into the preserve. The Washington State Department of Natural Resources manages the preserve, so a Discover Pass is required. There is an outhouse at the parking area, a 0.5-mile-long (800 m), wheelchair-accessible trail, a picnic area, and an interpretive display and viewpoint at the domed interpretive center.

To reach stop 2, backtrack to the junction with Maytown Road SW. Continue south on Mima Road SW for 1.3 miles (2.1 km) and turn right onto Bordeaux Road SW. Proceed 0.4 mile (0.6 km) and park at the gated gravel road. Do not block the gate. Spring is a great time to visit the preserve, when a sea of blue camas and other flowers bloom among the green grass.

6

Enigma on the Prairie

THE MIMA MOUNDS

The Mima ("mime-uh") mounds that dot the prairies of the southern Puget Lowland provide one of the most puzzling and contested natural history mysteries in the Northwest. These enigmatic mounds take their name from Mima Prairie, one of several natural grass prairies found in Thurston County between the towns of Tumwater, Rochester, and Yelm. Mounds are easy to observe; they cover about 10,000 acres (40,000 ha) and line miles of roads in rural Washington south of Olympia. Similar naturally occurring mounds are found in several regions of North America, from southern British Columbia to northern Mexico and as far east as Texas and western Louisiana. Hundreds of thousands of these close-set mounds dot prairies of the West. They are reported from as far away as the Pampas of Argentina. Many researchers restrict the term *Mima mounds* for those on the prairies of the southern Puget Lowland; *Mima-like mounds* is a catch-all for all others. At other mound sites, these low bumps are known as "prairie mounds," "hog wallows," "silt mounds," and, unfortunately, "pimple mounds." Mima-like mounds are found elsewhere in Washington; they are common in undisturbed areas of the eastern part of the state, where they are sometimes called "biscuits." Mounds are pretty quickly destroyed by agriculture and grazing.

The Mima mounds at the preserve are generally circular to elliptical, between 3 and 8 feet (1 and 2.4 m) high, and 8 to 60 feet (2.4 to 18 m) across. They are typically adjacent to each other, with as many as ten mounds per acre (0.4 ha). Large ones may contain 50 cubic yards (40 m³) of soil. They are separated by gravelly low spots, which have little to no soil cover. The interior of the Mima mounds consist of structureless (unlayered), black sandy loam containing scattered pebbles or even cobbles. The dark, mounded soil rests on top of stratified beds of gravel. The contact between the dark soil and the gravel may curve downward slightly into the gravel, giving the mound a curved upper and lower surface in cross section. The gravel is likely the source for at least a portion of the pebbles and

Mima mounds north of the observatory dome at Mima Mounds Natural Area Preserve.

cobbles in the mounds. It is outwash deposited by streams flowing out of the terminus of the Puget Lobe of the continental ice sheet 16,000 years ago, when the margin of the glacier was about 3 miles (5 km) north of Mima Prairie. The mounds are younger than the outwash, but whether by only a few years or a hundred or a thousand is debatable. Radiocarbon dates are equivocal, as charcoal found in the mounds has been mixed up by bioturbation. Whether this is due to animals burrowing into them and adding soil or displacement by roots, or both, plays a role in the origin debate.

Literally dozens of explanations have been put forward for the Mima mounds, ranging from the scientifically serious to the fantastical. So far the physical evidence has not produced a generally acceptable theory. There is a distinct controversy between biological and geological origin. A group of biologists have put forward an origin based on the actions of gophers; geologists are not united around any given hypothesis, but most, if not all I have surveyed, remain skeptical that rodents made the mounds.

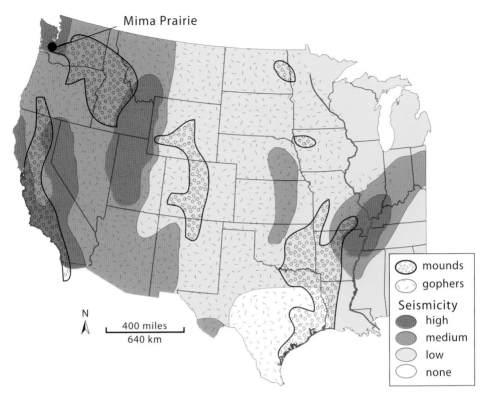

Mima Prairie

The distribution of Mima-like mounds, the ranges of eleven pocket gopher species (though they may not actually be present), and levels of potential seismicity. (Redrawn from Washburn 1988, Berg 1990, and Cox and Berg 1990.)

Local Native Americans, the Upper Chehalis, tell a story of a girl named Thrush who refused to wash her face fearing something awful would happen if she did. After she was finally convinced to clean up, it immediately began to rain buckets, flooding the land. The story holds that the Mima mounds are reminders of the waves that washed across the prairie. Charles Wilkes, of the US Exploring Expedition, examined the mounds in 1841. He thought they were Native American burial sites until the excavation of three showed they were not. He concluded they must have been man-made, perhaps for the cultivation of camas bulbs, a dietary staple of western Native Americans. James Graham Cooper, a prominent American naturalist working with a railway survey in the mid-1850s, suggested the mounds resulted from eddies and whirlpools (but didn't say how) from a time when the area was submerged by Puget Sound to the north, or that the region was part of a river estuary. The famous

Canadian landscape painter Paul Kane visited the Mima mounds in 1847. His painting Prairie de Bute *(Prairie of Mounds) is in the Royal Ontario Museum, Toronto.* —With permission of the Royal Ontario Museum © ROM

nineteenth-century ice age theorist Louis Agassiz concluded that the mounds were the nests of prehistoric suckerfish, based entirely on secondhand descriptions.

More rigorous investigations gave rise to several geologic hypotheses. These can be divided into depositional and erosional categories. In 1913, J Harlan Bretz, the pioneer glacial geologist of the Northwest, discarded the ideas previously bandied about and suggested something altogether new. Referred to as the sun cup hypothesis, he proposed that streams flowing over a thin, icy surface overlying the outwash concentrated mounds of sediment in pits melted into the ice by the sun. When the ice melted, the mounds of sediment were left. Bretz did not explain the origin of such a thin sheet of ice covering such a wide expanse. Though Bretz had doubts that this hypothesis could account for the sheer number of mounds at a range of elevations as well as on slopes, his idea was widely accepted. It was definitely better than Agassiz's suckerfish.

In 1940 R. C. Newcomb of the US Geological Survey put forward a different sort of ice and water hypothesis, this one erosional. It was considered an improvement on Bretz's, for a while. Newcomb saw frost polygons that form in polar regions as a model for the

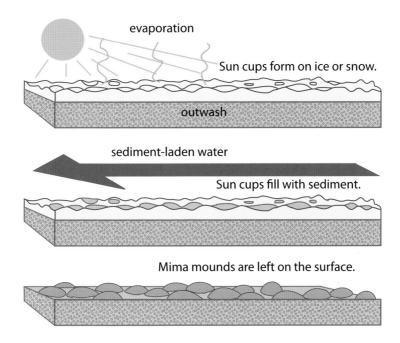

J Harlan Bretz's sun cup hypothesis. He did not propose a source for the sediment-laden water or account for the occurrence of Mima mounds at a variety of elevations.

mounds. Frost polygons form in permafrost-affected soils. Water-saturated soil is heaved upward by the expansion of water when it freezes; repeated freezing can increase the size of the piles by heaving the soil ever higher. Newcomb proposed that during the Pleistocene ice advances, soils in the periglacial, or near-ice, environment froze repeatedly, resulting in the mounds. An objection to this idea was that frost polygons aren't so named without reason: how did polygonal frost heaves become circular mounds?

In 1953 Arthur Ritchie, a geologist with the state highway commission, tried to answer that flaw in Newcomb's idea. He posited that thawing polygonal blocks were subjected to meltwater erosion, which rounded off their edges. Alas, this sort of erosion is not observed in the arctic environment during seasonal thaws. Also, there is no solid evidence that permafrost existed in the vicinity of the Vashon glacier in the Puget Lowland. Annual temperatures during the Vashon advance are estimated to have been 11°F (6°C) cooler than modern conditions in western Washington, but even still the temperatures would have been above freezing on average annually. However, cold katabatic winds flowing off the ice could have

Frost polygons in the McMurdo Dry Valleys of Antarctica.
—Courtesy of David Marchant, National Science Foundation

Ice forms between polygonal blocks of frozen silt.

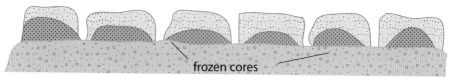

frozen outwash gravel

Polygonal blocks form in frozen silt overlying outwash gravel.

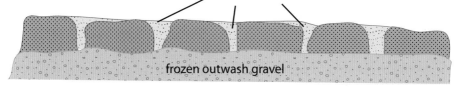

frozen cores

Ice wedges and surface of blocks melt; cores and gravel remain frozen.

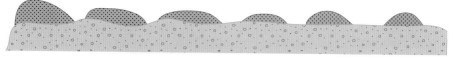

Floods wash away thawed sediment, which later thaws as mounds.

Arthur Ritchie's idea of how the erosion of frost polygons could have formed the Mima mounds. (Adapted from Jackson 1956.)

depressed local temperatures near the ice, so permafrost may have developed in the vicinity of the Puget Lowland.

Biologists entered the fray in 1942, when prominent vertebrate biologists Victor Scheffer and Walter W. Dalquest presented the gopher hypothesis, which posits that burrowing pocket gophers built the mounds as nesting areas. At first glance, this depositional hypothesis sounds outlandish: how could small rodents weighing as little as 2 ounces (60 grams) construct the six thousand or more mounds found just in the Mima Mounds Natural Area Preserve?

Scheffer and Dalquest figured that these mammals migrated northward as the ice receded. Soils were very young and thin above the outwash. The gophers dug deeply into the gravels below the soil to build nesting cavities and carried excavated soil to the surface. Biologists who favor this hypothesis contend that this was the start of the mounds. Once mounds had developed, gophers preferentially occupied them to escape flooding or saturated soils, enlarging them by scraping soil onto the mound from the prairie surface. This theory holds that the intermound areas, floored with gravel today, are the areas where the gophers scraped away the soil.

Many biologists hold 2- to 3-ounce (60 to 85 g) pocket gophers responsible for building the Mima mounds. The Mazama pocket gopher (Thomomys mazama) *of Thurston, Pierce, Clark, and Mason Counties is listed as a threatened species by the state of Washington.* —Courtesy of ©Ty Smedes

Pocket gophers don't live on Mima Prairie today (they do live on other Puget Lowland prairies), and scientists can find no direct evidence that they ever did. Still, a number of arguments favor the gopher hypothesis. Mima-like mounds are found only where pocket gophers are known to live, or, according to gopherists, may have lived at one time, which is most everywhere west of the Mississippi. The spacing of the Mima mounds closely resembles the average distance between burrows of these antisocial animals. A single family of pocket gophers can move 5 tons (4,500 kg) of soil in a year, or about one-twentieth of the mass of a typical Mima mound. With this information in mind, a 2014 study using a computer model showed that generations of (virtual) gophers could build mounds over centuries. And then there are the mysterious "mound roots" found in some excavated mounds. These 0.5- to 2-foot-wide (15 to 60 cm) extensions of soil penetrate the underlying gravel. Gopherists emphasize that the mound roots are about the size and shape of the nest burrows and tunnels dug by pocket gophers.

Biologists who are skeptical point out that pocket gophers build much smaller mounds, at most about 6 inches (16 cm) tall, and these only at the mouths of burrows, not scraped off the surface. Some gopherists retort that the Mima mounds of western Washington occur in areas of well-drained thin soil over gravel, and it's the presence of the gravel that accounts for the otherwise anomalous mounding behavior. They admit that in areas where the soil is deep, these critters may not pile soil into mounds at all, and certainly none that are anywhere near the size of the Mima mounds. Other biologists poke holes in the gopher theory, arguing that busy gophers moving soil on the surface are sitting ducks for raptors. Also, stones measuring up to 20 inches (50 cm) across are present in the mounds; several biologists point out that these are far too large for little rodents to move. Perhaps the most serious argument against the gopher hypothesis is that burrows, when found in Mima mound prairies, are between the mounds rather than on or in them. Nonetheless, the gopher hypothesis is widely favored by biologists.

The ability of vegetation to anchor soil from wind or water erosion forms the backbone of the runoff erosion–root anchor hypothesis. This idea was introduced by George Gibbs, the pioneer of geologic research in the Washington Territory back in 1854, and has been fine-tuned by several researchers since. The hypothesis holds that in the early Holocene, after the Pleistocene ice had receded, clumps of bushes and, later, trees became established on thin soils, which originated as windblown dust deposited near the glacial margin. Their roots anchored the light, silty soil. Subsequent small-scale

A cartoon by famed cartographer Dee Molenaar, illustrating how gophers may have constructed the Mima mounds. —Courtesy of Dee Molenaar

stream erosion removed soil between the clumps of bushes, beyond the influence of roots, leaving the gravel-paved intermound depressions. Proponents of this idea point to soil mixing by roots to account for the presence of stones within the mounds.

Albert Lincoln Washburn was a highly regarded arctic geomorphologist at the University of Washington. His description of the Mima mounds, published in 1988, included an exhaustive examination of the leading hypotheses in an effort to sort out the controversy.

He determined root anchoring to be the most likely among them due to observations that western Washington's Mima mounds, at least, are found only in alluvial areas, meaning running water was present, and that the mound prairies are confined to terraces built during the deposition of the Puget Lobe outwash. This geomorphic association provides a temporal connection between the Mima mounds themselves and the years just following the demise of the Vashon glacier, when vegetation was establishing itself on the young soils developing on the outwash plain.

Washburn maintained that Mima-like mounds found elsewhere in North America could have a variety of origins, since most are found nowhere near environments that were adjacent to glaciers during the Pleistocene. His major disagreement with gopherists revolved around the presence of the large cobbles in the mounds.

Others dispute Washburn's conclusion, pointing out that if the mounds had formed soon after the glaciers melted, they should be much reduced in size due to progressive slope degradation caused by wind and water erosion, bioturbation, rainstorms, and slumping. In their view, the mounds are much younger than Washburn's model requires. At least one biologist, however, points out that the mounds are, at least today, well covered with prairie grasses with tough root mats, which would reduce the effects of erosion.

Geologists have introduced new hypotheses since Washburn's analysis. Andrew Berg, a geologist with the US Bureau of Mines in Spokane, shook things up with an altogether new idea in 1990. The seismic shaking model is based on personal observations he made while building a doghouse in his backyard. Shortly after the 1980 Mount St. Helens eruption, Berg was hammering on a board covered with a fine layer of volcanic ash. He noticed that the fine grit tended to mound up into piles when the board was struck. He hypothesized that seismic waves caused by violent earthquakes caused thin, low-density sediment (soil) resting on a hard substrate (the gravels) to jump upward due to interference patterns between the original seismic waves and those reflected off rock layers of different densities. The soil came to rest in areas where interfering waves canceled out each other's relative up or down motion. Berg also noted that Mima-like mounds, at least in the United States, correlate well with seismic areas and poorly with areas where gophers live today. Berg does not account for vegetation anchoring the soil and inhibiting the piling effect. For his theory to hold water, the

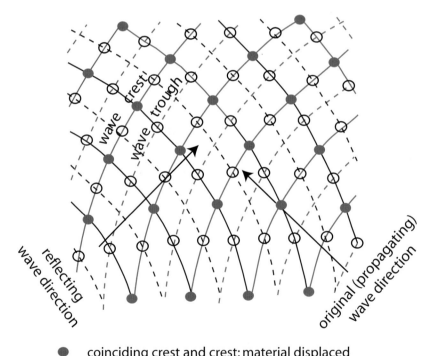

● coinciding crest and crest: material displaced
○ coinciding crest and trough: material concentrates

The interference pattern of intersecting seismic waves. Loose soil particles located at the red dots would be displaced and bounced into the air. Soil at the open circles would not, due to the cancellation of wave effects. Mounds would form at these intersections. (Redrawn from Berg 1990.)

earthquake or earthquakes must have happened just before vegetation took hold on dust- or soil-covered glacial outwash. Vegetation must have appeared right after the earthquakes, before erosion
could degrade the mounds.

In a 2009 study, Robert Logan and Timothy Walsh, geologists with
the Washington State Department of Natural Resources, used lidar
imagery to locate Mima mounds that may not be obvious on the
ground. Lidar uses an airborne laser to image the ground surface,
seeing right through the obscuring fuzz of vegetation. The researchers found that Mima mounds south of Puget Sound, some of which
are well covered by forests, closely correspond to the former margins of the Vashon glacier and are all located in glacial outwash
channels. They postulate that water bursting out of the stagnant,
rapidly melting glacier formed the outwash channels. This idea is
not controversial. They go on to postulate that water ponded in the
outwash channels, perhaps dammed by chunks of ice, and froze.
These sheets of ice were then pitted with cuplike depressions as the
ice melted; sun cups are common in springtime mountain snow-
packs today. Subsequent floods bursting out of the thinning and collapsing glacier filled these cups with sediment, which was left on
the outwash plains after the ice melted.

Logan and Walsh admit that this rehashes J Harlan Bretz's 1913
idea, which Bretz himself had a hard time with. There are a number
of weaknesses in the sun cup hypothesis. One of the most glaring
is that the sediment deposited in the sun cups was remarkably fine
grained and well sorted, textures that are not usually associated
with deposits left by glacial outburst floods; these deposits tend to
be coarse and unsorted. And there are mounds on outwash channels of different elevations and, consequently, different ages. Some-
how the older mounds managed to survive the outburst floods that
led to the formation of the younger mounds. It seems unlikely this
could be the case.

Though the mounds remain a mystery, the source of their sediment is no longer so. A portion of the sediment, including the cobbles in the black soil, is andesite from Mount Rainier. The Puget
Lowland prairies lie along what is believed to be the course of a catastrophic debris flow that occurred 16,000 years ago. It was released
when an ice dam across the mouth of the Nisqually River collapsed.
The receding Vashon glacier had dammed the river mouth, creating a large lake in the valley. The debris flow deposit weathered to
become the soil of which the mounds are made.

With your head full of conflicting theories, go to stop 1 to see
these puzzling mounds up close. From the parking lot, take the

paved trail out into the remarkable, nearly treeless prairie. Be sure to stop at the squat concrete dome of the interpretive center and take in an overview of the mounded Mima Prairie from the raised viewpoint. There are numerous gravel trails from which to explore the prairie as well, contemplating a geologic versus biologic origin for the mounds. There are plenty of molehills visible, too. The moles bring quite a bit of gravel to the surface and serve as an example of bioturbation.

Proceed to stop 2 to see a cutaway of a Mima mound. Duck under the gate and walk east along the barbed wire fence lining Bordeaux Road SW for about 200 feet (65 m), until you come to a small sign (unrelated to the Mima mounds). Use care, because there is no path; depending on the time of year, you may be walking in knee-high grass. Turn away from the road and walk across the grassy prairie about 150 feet (45 m) to a Mima mound that the Department of Natural Resources excavated with a backhoe. The contact between the black silt of the sectioned mound and the underlying gravel is knife sharp. You may see a number of mound root structures protruding downward into the gravel, though they could erode over time. The silt is rich in pebbles; those larger than 2 inches (5 cm) argue against

"Whence came ye?" Pebbles stud the dissected mound at stop 2. Several 6- to 8-inch-long (15 to 20 cm), dark mound roots protrude down into the outwash gravel. Are these a clue to the formation of the Mima mounds?

the pocket gopher hypothesis, but pebbles under that size do not. The outwash gravel contains plenty of larger pebbles.

The Mima mounds have generated more hypotheses for their origins than the average geologic feature, and they will likely continue to do so. Every now and then newspapers and the Internet carry a headline akin to "Mima Mound Mystery Solved!" Take that with a shovel or two of salt. Personally, I'm going with the Paul Bunyan hypothesis, which has at least two variants.

In one, the legendary logger wanted to transport logs to mills more easily. So ol' Paul dug out Puget Sound and dumped the dirt on the prairies to the south. Another version tells us that Paul Bunyan heard about the Great Wall of China and wanted to raise such a wall in the United States. The workers he hired to dig dirt for the wall became upset because the job seemed endless. They went on strike and left their wheelbarrows on the Puget prairies. After the wooden wheelbarrows rotted, the remaining mounds of earth became the Mima mounds. Before you leave the Mima mounds, you may want to take some time to come up with your own theory or decide if an already published one makes more sense than the others. The absence of a general theory just adds to the wonder of these mounds and serves as inducement to visit again and again, in all seasons.

Paul did it. This 31-foot-tall statue of the famous lumberjack is in Kenton, a neighborhood in northern Portland, Oregon.

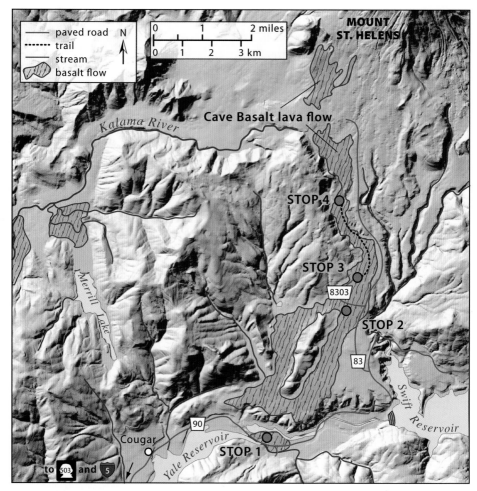

Younger lahars and pyroclastic flows cover parts of the Cave Basalt lava flow, and patches of it appear isolated in the Kalama River valley to the west. (Geology from Greeley and Hyde 1972.)

GETTING THERE: From exit 21 on I-5 in Woodland, head east on Washington 503 (Lewis River Road), which eventually becomes FS Road 90, for 28.8 miles (46.3 km), passing through the village of Cougar. Stop 1 is a roadside pullout 2.4 miles (3.9 km) east of Cougar.

Continue east on FS Road 90 and turn left onto FS Road 83, about 35 miles (56 km) from I-5. After 1.8 miles (2.9 km) turn left onto FS Road 8303. After 0.2 mile (0.3 km) the road reaches stop 2, the Trail of Two Forests. It is another 0.7 mile (1.1 km) on FS 8303 to the large Ape Cave parking area, stop 3. Stop 4 is the 1.5-mile (2.4 km) trail connecting the two entrances to Ape Cave. Stops 2 through 4 are in Mount St. Helens National Volcanic Monument, so collecting rock samples is prohibited. Parking passes are required at these stops as well. All roads are paved.

7

Mount St. Helens National Volcanic Monument

The basaltic lava flow that hosts Ape Cave and the Trail of Two Forests is formally named the Cave Basalt. This lava flow is one of only two basaltic lavas mapped at Mount St. Helens, which otherwise is built of large volumes of andesite and dacite. The flow, erupted from a vent somewhere around 5,000 feet (1,525 m) in elevation on the mountain's southern flank, is about 8 miles (13 km) long. The vent and uppermost part of the Cave Basalt is buried by a younger lava flow, the 350-to-450-year-old Swift Creek andesite, and lahar deposits. The Cave Basalt flowed down a stream valley and reached the Lewis River in the vicinity of stop 1. Another lobe of the basalt headed west, descending the upper Kalama River valley for about 6 miles (10 km) and damming a tributary to form Merrill Lake. The Cave Basalt erupted about 1,900 years ago.

The Cave Basalt is a pahoehoe-style lava flow. *Pahoehoe* (pronounced pa-hoi-hoi) is a Hawaiian word derived from *hoe*, which means "to paddle." V-shaped patterns on pahoehoe lava reminded Polynesians of the swirled wake of a canoe in the sea. Pahoehoe is characterized by a relatively smooth surface, often with convoluted and beautiful braided shapes. Pahoehoe lavas have relatively low viscosity, meaning they are fluid or runny and can advance laterally across essentially flat ground. Pahoehoe flows typically spread as

Ape Cave is very popular, especially with busloads of schoolkids. Up to 135,000 people visit the lava tube per year. Time your trip accordingly. Dry weather in the late fall to early spring is optimal. Most people walk the lower part of the cave, making the round-trip in an hour or two. Budget at least two hours in the lower cave, another three in the upper, and at least an hour for the return hike on the surface trail. If you wish to visit all four stops, plan to spend the entire day.

thin lobes about 3 inches (8 cm) thick. The lobes thicken as lava continues to erupt and flow within a hardening crust, inflating the flow with pressurized lava like a balloon. Flows are often no more than 10 feet (3 m) thick if unconfined by topography, although some particularly large-volume flows, such as those of the Columbia River Basalt Group (vignette 5), are far thicker. Pahoehoe basalt temperatures at the time of eruption are between 2,000°F and 2,200°F (1,100°C and 1,200°C).

Stop 1 is near the terminus of the Cave Basalt. Beginning at the Swift #2 hydropower-generation station east of Cougar, the Cave Basalt extends eastward along the road for about 1 mile (1.6 km). The vegetation changes abruptly once you reach the lava, as the soil is rocky and thin on top of the young lava flow. The forest consists of stunted trees, with lichen and mosses growing on the bare rock. Several decades if not hundreds of years passed before sufficient soil developed on the hard lava pavement to support a forest. A gravel road branches to the right 0.8 mile (1.3 km) east of the hydrostation and is a convenient place to park.

At stop 1, a freshly broken surface shows that the Cave Basalt is riddled with gas pockets, or vesicles. The 0.04- to 0.08-inch (1 to 2 mm) white minerals are plagioclase feldspar. Green olivine is common but too small to see in the photo.

The Cave Basalt here is typical of the entire flow. The surface is weathered and crusted with a thin coating of lichen and moss, so use a rock hammer to break off a piece to examine a fresh surface. This stretch of road is outside the national monument, so you can whack on the rock without risking going to jail. The rock contains a lot of vesicles, up to 0.4 inch (1 cm) across, and is rich in crystals. Clear to white plagioclase predominates, but with a hand lens you can spot bottle-green olivine.

Drive down the gravel road, cross the narrow bridge over the Lewis River, and turn around. The Cave Basalt flowed into the river here. The 15-foot-high (4.5 m) wall of rock exposed on the northern bank is Cave Basalt, but the rock in the river on the south side is much older lava. The contact between the older lava and the thin leading edge of Cave Basalt was eroded away by the river.

Trail of Two Forests

Stop 2 is the Trail of Two Forests, a wheelchair-accessible, 0.25-mile-long (0.4 km) boardwalk on the Cave Basalt. The holes, or trunk molds, dotting the lava are the main attraction. As lava advanced through the forest that existed here, it was quenched against the trunk of each tree; even burning wood is cold compared to lava. The lava hardened and formed a rocky seal around each tree trunk and kept lava out of the space left behind as the trees burned. The well-like holes left behind are as deep as the lava flow and preserve the diameter of the tree trunks. With a flashlight you can examine the inner surface of a mold. In some you can see the pattern of charred wood in the lava. A new forest now grows on the surface of the lava.

Another major attraction is the 3-foot-wide (0.9 m), 50-foot-long (15 m) tunnel called the Crawl, the hollow mold of a log the lava flowed over and burned. You can crawl through this mold by descending a short metal ladder.

From the boardwalk you can see the intricate patterns on the lava surface, including the classic signature of pahoehoe: braided rope-like coils of basalt. Wrinkled-looking lava such as this is called ropy pahoehoe. It forms as the thin solidifying crust of the lava slows down and is less able to flow. Drag from the warmer lava flowing beneath the still-flexible crust wrinkles and deforms it into fascinating folds. As the lava cools and forms a thicker crust, the crust becomes more viscous, until ropes no longer form. At that point the solidifying lava on the surface tends to break into slabs and chunks rather than smoothly deform into these beautiful shapes. There is a nice tilted slab of ropy pahoehoe on the opposite side of the parking lot from the start of the boardwalk.

The charred surface of a log is clearly preserved in the basalt lava of the Crawl.

An incandescent toe of pahoehoe lava oozes onto pavement at Kilauea in Hawaii. —Courtesy of ©Ron Schott

Solidified ropy pahoehoe at the upper entrance to Ape Cave.

Ape Cave

The Ape Cave lava tube, stop 3, formed in the interior of the Cave Basalt as the lava flowed down a stream valley. The lava tube is divided into two sections: The lower part is a round-trip stroll of about 1.5 miles (2.4 km). The upper part is 1.5 miles (2.4 km) one-way and a much rougher trip.

Ape Cave is one of the most popular geologic attractions in Washington. Every year, tens of thousands of people descend the iron stairway into the dark subway-like lava tube. Lava tubes form in basalt lava flows that are runny and flow down a slope. A gradient as low as 1° is steep enough. Ape Cave is widely held to be the longest lava tube in the world at 11,330 feet (4,453 m), but this is far from true. While the cave is the longest continuous lava tube in the continental United States, there are many much longer lava tubes. The Kazumura lava tube complex, on the east flank of Kilauea on Hawaii, is the longest; it is nearly 36 miles long (58 km).

Sometime in the late 1940s or early 1950s, Lawrence Johnson, a logger, discovered the cave and reported it to Harry Reese, an avid caver and storeowner at Cougar. Reese sponsored a local Boy Scout troop called the Mount St. Helens Apes. They took their name from a famous tale told by a party of miners. The miners claimed that one night in 1924, while holed up in a cabin on the east flank of the mountain, stone-throwing "hairy apes" attacked them. The miners really were attacked, but by two local jokers (one of them confessed in 1982), who managed to thoroughly scare the miners. This tale led to the association of apes with Mount St. Helens, then to stories of Sasquatches living in caves, and eventually to the cave-exploring Mount St. Helens Apes. The Cascade Grotto of the National Speleological Society mapped the tube in 1978.

When low-viscosity lava erupts, it obeys gravity and flows down valley bottoms. The bottom and edges of the flow cool and crust

PROTECTING APE CAVE

Careless explorers damage cave formations. Once a cave is damaged, it remains damaged forever. For this reason, please observe these common sense regulations:

- No food, beverages, or litter. These items attract animals and bacteria that do not belong in the cave. Food and litter decompose at an extremely slow rate in the cool temperatures of caves.

- No smoking. Smoke lingers in the cave and is harmful to cave creatures and humans.

- No flares, fireworks, firearms, or any kind of open flame. These leave a residue and blacken cave walls.

- No rock collecting or damaging cave features (steep fines).

- No pets. Animal feces damage cave ecosystems and are unpleasant for other visitors.

- Do not touch the walls. Bacteria with fungus-like characteristics, called "cave slime," live on cave walls. This slime is a food source for other cave life, is easily wiped off when touched, and takes many years to regrow.

over from contact with the much colder ground. The upper skin cools more gradually but eventually crusts over. The interior of the lava flow is still molten and is provided a continual recharge of lava at the vent, pressurizing the flow within its crusted margins. Flowing lava is concentrated in the center of the flow. Eventually, as cooling migrates inward, lava movement is restricted to long distributary tubes inside the hardening flow (think of a garden hose full of water). Once confined within a tube, fluid basalt may cool at less than about 1.8°F (1°C) each 0.6 mile (1 km) of distance it travels, according to measurements from Hawaiian lavas. Eventually the eruption of lava decreases at the vent. The terminus of the lava continues to advance downhill as long as there is molten lava for gravity to act on. The tubes that supplied lava to the leading edge may drain and empty as the advance continues, leaving hollow tunnels behind (the empty garden hose). Ape Cave is such a drained lava tube. Hollow tubes may serve as passages for later flows. This was the case at Ape Cave.

The cave entrance, a large skylight resulting from the collapse of the tube's ceiling, is at about 2,090 feet (640 m) of elevation. Lawrence Johnson made his discovery here. A metal stairway goes down into the gloom. At the bottom the ceiling is about 40 feet (12 m) high. Here you have a choice of going up-tube (back under the stairway) or down-tube. I recommend exploring the lower section before tackling the wilder upper part.

Your first impression will be of the immense size of the lava tube; it resembles a subway tunnel more than anything else. Lava

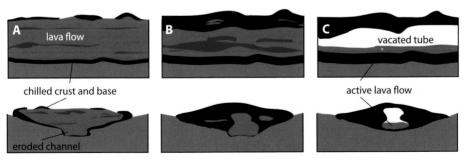

Longitudinal view (top) and cross sections (bottom) showing the three stages of tube formation. (A) Lava flows down a valley. The lava flow's outer edges and some of the surface cool and solidify. Once the flow is thick enough, it begins eroding downward into soft substrate. Lava becomes confined to tubes within the flow (B) as cooling expands through the flow. As the eruption wanes, lava drains out of the interior at the flow's terminus, leaving a hollow tube behind (C). This drawing is not to scale; lava tubes may be quite small compared to the cross section of an entire flow. (Adapted from Greeley and Hyde 1972.)

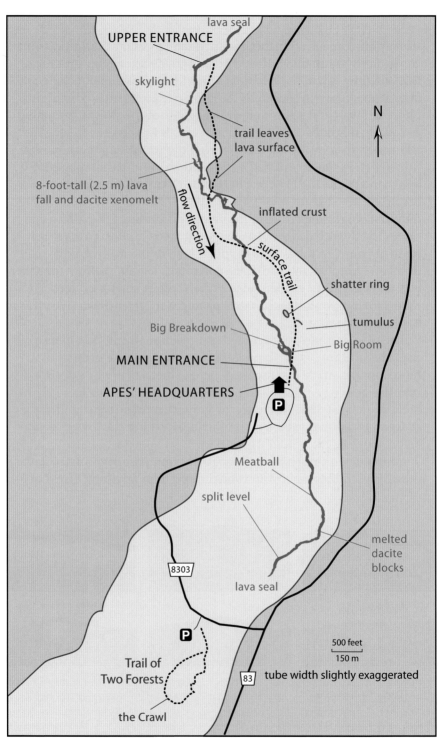

Map of the Cave Basalt flow (purple), Ape Cave (red), cave features (red labels), the surface trail (dashed line), and features along the trail (green labels). (Modified from Greeley and Hyde 1972 and Halliday 1983.)

between the cave's ceiling and the surface may be up to 30 feet (10 m) thick, but in most places it is closer to 15 feet (4.5 m).

The whole tube was once filled with flowing lava. A series of subtle parallel lines are often visible in the walls of Ape Cave. They were eroded into the softened cave walls by blocks carried in the lava. Ledges protrude from the walls, marking temporary high lava levels in the tube. The lava drained out of the tube, some of it reaching beyond stop 1, a little more than 4 miles (7 km) south of the main entrance. A small lava flow covers the floor for much of the first 300 feet (90 m) of lower Ape Cave (it extends far up into the upper cave). This flow occurred during a late stage of the Cave Basalt eruption, entering the empty tube after the main flow had exited. It left raised, parallel walls, or levees, along the young flow's outer margin as it was chilled against the floor and walls of Ape Cave. The remainder of the small flow drained downslope between the levees.

The most famous feature in the lower cave is the Meatball. It is one of several blocks of cooled lava that likely fell from the ceiling

Parallel bathtub rings line the walls of Ape Cave for most of its length. Levees along the edges of a small lava flow that came later form "railroad tracks" on the floor. —Courtesy of John Scurlock

In places, the cave walls narrow toward the top. —Courtesy of John Scurlock

while lava was still flowing. It was carried down the tube, becoming coated with lava and rounded. It wedged in a narrow spot where the flow was forming an interior roof about 12 feet (4 m) above the present cave floor.

Studies at Ape Cave provided the first unequivocal evidence that lava flows mechanically and thermally erode the deposits they flow over. Most basalt flows are part of extensive flow fields, and they typically flow over the surface of older flows. Evidence for erosion into the solid rock of older flows is scanty because the old and new lavas usually have very similar, if not identical, chemistry and the same melting points. The Cave Basalt is different. It flowed down a stream valley filled with unconsolidated sediment of a lahar deposit, composed mostly of blocks of dacite, the most typical rock erupted from Mount St. Helens. Dacite has a demonstrably different chemistry from basalt, principally in the proportion of silica, and erupts at much lower temperatures, so it has a lower melting point. The fast-moving basalt lava eroded dacite blocks, at least partially melted them, and incorporated them into the lava flow. Chemical variations in the basalt and the visibly melted (or at least thermally

softened) dacite rock fragments (called xenomelts) in Ape Cave provide evidence that the basalt thermally and mechanically eroded
the substrate it flowed over.

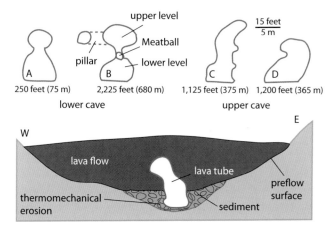

*Cross sections (A through D) through Ape Cave looking upslope. Lava
tubes have complex shapes due to flow dynamics, thermal erosion, and
postflow ceiling collapse. Top: The distances in feet are from the main entrance; A and B are below it, and C and D above. A is just below the main
entrance; B is at the Meatball; C is at the Big Breakdown (ceiling collapse
has created a high cupola); and D is just beyond the first breakdown. Bottom: lava thermally eroded a channel downward into dacite boulders of a
pre-lava streambed. (After Hyde and Greeley 1973.)*

A couple of gentle curves and about 200 feet (60 m) beyond the
Meatball, watch for a solitary 3-foot-high (1 m) fallen block in the
middle of the passage; it serves as a landmark for the next point of
interest. Behind another block along the right-hand wall the glassy
wall has broken down and exposed a cross section of the remarkably
thin lining of the tube—no more than 6 inches (15 cm) thick—and a
shallow cavity beyond. Stick your head in there with a bright light
and examine the blocks of rock outside the lava tube's wall. These
are in the lahar deposit the lava burrowed into and melted. Many of
the blocks, especially the reddish ones, are spiny and rough, but the
yellowish ones are softer looking, having been thermally deformed.
Some of the yellow rock oozed between other blocks or even drib-
bled onto the floor of the cavity. The yellow blocks are dacite; the
oxidized red ones are most likely andesite, which has a melting
point closer to basalt. Once solidified, the surprisingly thin walls of
the lava tube were sufficient to insulate the flowing lava from the
relatively cool rocks outside the walls. These thin walls are a good

illustration of the ability of rock to insulate. There are other places throughout the cave where the tube walls have collapsed to expose thermally deformed rocks; another is down the tube from this one, on the left.

Near the end of Ape Cave, the high, wide passage suddenly narrows, and the ceiling lowers to form two tubes, one stacked above the one you are in. The upper one is not a separate tube. It was separated from the lower part of the tube when the lava flowing through the lower tube crusted over. When it drained out of the tube, it left the ceiling. It is difficult to get onto the upper level, which in any case quickly ends.

You may notice that the floor of Ape Cave is now covered with up to 3 feet (1 m) of gray sand and white pumice lapilli, swept into the cave when a lahar flowed into the cave's main entrance. The sediment is ash and pumice of the W tephra, erupted from Mount St. Helens between 1480 and 1482 CE. It may have been emplaced at

The glassy lining of the lava tube was blasted inward by trapped gases. The oxidized, unconsolidated red sediment the Cave Basalt burrowed into is exposed behind. Molten basalt dripped or flowed downward (right) to form what's called wall drapery.

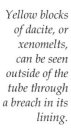

Yellow blocks of dacite, or xenomelts, can be seen outside of the tube through a breach in its lining.

that time or during a later lahar. Some of the lahar deposit rests on ledges 2 feet (0.6 m) above the cave floor. It would not be good to be in Ape Cave when an event like this recurs!

The ceiling gets lower and lower; eventually you have to duck-walk and then crawl until the tube is completely sealed off. This end of the lava tube is about 2,000 feet (600 m) southeast of Apes' Headquarters.

Upper Ape Cave

The upper, more rugged section of Ape Cave starts at the main entrance stairway. At the bottom of the stairs, turn sharply and walk back under them. The upper cave is considerably more difficult to traverse. Occasional high piles of fallen rock are the main obstacles.

These breakdown piles have fallen from the cave ceiling; one pile nearly reaches the ceiling. Negotiating a breakdown pile requires stepping on large lava blocks and over the gaps between them. The other major obstacle is a vertical lava fall that rises about 8 feet (2.5 m). Once you have gotten over that, it is pretty clear sailing all the way to the upper entrance of Ape Cave at 2,460 feet (750 m) of elevation. The upper section is about 1.5 miles (2.5 km) long and gains about 370 vertical feet (110 m).

Walking away from the stairs, you'll quickly come upon well-formed levees on the floor left by a small lava flow that entered Ape Cave after the main Cave Basalt flow had drained out of the tube. In one place the top surface of the small flow crusted over and formed its own mini–lava tube. The roof of this tube is broken open to reveal this tube-in-tube feature. Watch for fallen blocks of lava that fell before the floor had solidified; they appear to sink into the floor and may have signs of flow around them. You may also find subtle ripple marks where a block fell in and sank.

About 300 feet (90 m) from the stairs you enter the Big Room, formed as blocks fell from the tube's ceiling to form a cupola. A cupola can migrate upward into the solid lava above the tube; some

A breakdown pile nearly plugs upper Ape Cave at the Big Room. The floor is covered with very smooth lava. —Courtesy of John Scurlock

cupolas may begin to form while lava is still flowing in the tube, and they may continue to migrate upward via rock fractures long after the flow has solidified. If the collapse is extensive, the cupola may reach the surface, breaking through to form a skylight. The basalt above the Big Room is estimated to be 21 feet (6 m) thick.

At the Big Room you'll encounter the 700-foot-long (215 m) Big Breakdown, a collapse pile that nearly fills the cave. The easiest route around is between the pile and the left side of the cave, passing by the opening to a short, low-ceilinged, 75-foot-long (25 m) tributary tube that drained into the main Ape Cave tube. If you climb over the top of the heap, be careful of loose blocks, and don't bump your head on the ceiling! There are more than a dozen breakdowns along the length of the upper tube, but none as large as the Big Breakdown.

Just beyond the Big Breakdown the cave floor is covered by rough-textured aa. *Aa* is another Hawaiian term, meaning "fire." Spiny blocks on the surface characterize this type of flow. There are no ropy structures. Pahoehoe and aa are the same rock type, but aa develops in a more viscous flow that is cooler or a more turbulent flow that fractures as it moves. A low-viscosity pahoehoe flow can become an aa flow, but aa cannot revert to pahoehoe; it is too fractured and viscous.

There are a few places where the cave wall was explosively blown into the lava tube. Red oxidized sand and small blocks are scattered around these openings. One explanation for these features is that confined steam generated from water or snow just outside the tube lining expanded quickly. An alternative is that vegetation buried by the lava generated hydrocarbon gases, which eventually exploded.

For most visitors to the upper cave, the most memorable feature is the vertical, 8-foot-tall (2.5 m) lava fall. Lava backed up behind the narrow section of cave here, perhaps blocked by a dam of fallen blocks. It eventually overflowed the obstacle to create the fall.

About 3 feet (1 m) below the falls, on the wall to the right, is a geologically important feature. A 6-foot-long (2 m) tongue of yellowish lava emanated from a hole and ran down the wall. This is xenomelt derived from solid dacite boulders in the pre–Cave Basalt lava stream channel. Melted blocks like this provided the first tangible evidence that lava flows can thermomechanically (through melting and physical force) erode their own channels. There are more examples of xenomelt beyond the falls. There is a toehold in the left rock wall and solid rock knobs on top of the wall to help you slither over the fall. A shove or a hand from your caving buddies will make it easier. I suggest you let the tallest person, or best climber, go up last!

The vertical lava fall in upper Ape Cave. —Courtesy of John Scurlock

This dacite xenomelt ran down the tube lining. —Courtesy of John Scurlock

Beyond the lava fall light from the outside world pours through a skylight, where a cupola reached the surface 30 feet (10 m) above. If this happened while the lava was flowing, a person standing on the hot, crusted surface could look down into the incandescent interior of the flow, feel the searing heat, and gag on the sulfurous fumes; an exciting but not overly pleasant experience. Some people have tried to exit the cave here and been seriously injured. The true exit is only 250 feet (75 m) farther, at a second skylight with a metal ladder to the surface. The cave ends about 450 feet (140 m) beyond the ladder.

Cave Basalt Lava Flow Trail

Stop 4 is the entire length of the wide, easy, 3-mile (4.8 km) round-trip hike connecting Ape Cave's two entrances. It begins near the west margin of the Cave Basalt flow, at the main entrance to Ape Cave (elevation 2,090 feet, 637 m), and wanders through the thin forest on the surface of the flow. The turnaround point is at the exit from the upper section of the lava tube (2,660 feet, 811 m). This description assumes you start at the main cave access and hike to the upper entrance and back. Read this section in reverse if you have traversed the upper cave and are heading down the trail. If you intend to return via the upper cave, you will have to climb down the 8-foot (2.5 m) vertical lava fall described in the preceding section. (See the appendix for GPS coordinates of the features along the trail.)

The first point of interest is a small stream crossing 0.1 mile (0.2 km) from the main cave entrance. A 1-foot-thick (1 m) sandy deposit of ash and white pumice from the W tephra rests on rippled pahoehoe. This may be from the same lahar that left similar deposits in Ape Cave. Lapilli-sized pumice of the W tephra is common on the surface around Mount St. Helens and is easily remobilized in running water.

Beyond this the trail crosses a flat sandy expanse with very little exposed lava; the lahar filled in the basalt topography. In 300 feet (90 m) or so the trail breaks out of the scrubby forest into a large opening in the middle of the flow. Look for a faint boot-beaten track branching right, which leads to the large, tilted slabs of a tumulus. Tumuli are common on pahoehoe flows. This one is about 170 feet (50 m) across. A tumulus forms when lava flowing beneath the crusted surface exerts upward pressure and fractures the rock above, perhaps because eruption volume increased at the vent or the flow was obstructed downstream of the tumulus. This shoves slabs of the crust up and sideways. Small volumes of lava may leak from a tumulus and flow across the lava surface.

In about 0.2 mile (0.3 km) the trail skirts a wall of lava blocks. A shatter ring, a semicircular structure with a relatively smooth lava floor, is visible from the rim of the wall. Circular or elliptical shatter rings form above active lava tubes. There are several on the surface of the Cave Basalt; this one, 20 feet (6 m) high and 250 feet (75 m) long, may be the most easily accessible example in the United States.

An aerial view of a shatter ring forming in Hawaii. The inner portion of the ring is collapsing into the tube beneath, but subsided blocks still cover the tube. A skylight has yet to develop here. The fresh, glassy, silvery surface is typical of pahoehoe. —Courtesy of Tim Orr, US Geological Survey

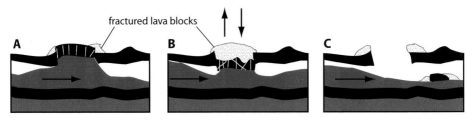

Stages in the formation of a shatter ring. (A) An increased volume of lava pulses down the tube and uplifts a thin slab of crust at a weak point. (B) The uplifted slab collapses as the pulse of lava passes. Repeated pulses lift the slab like a piston and fracture it into blocks. (C) After the lava in the tube subsides, the fractured crust may fall into the tube, with a ring of blocks remaining on the surface; if collapse is incomplete, a pit floored with blocks may remain on the surface. Fluid lava from within the tube or from a source outside the tube may then pond in the ring, cooling to form a smooth surface. (Based on a drawing in Orr 2011.)

The trail continues along the eastern edge of the flow; more-mature forest and dense brush cover the hillside on your right as you hike up the trail. About 0.25 mile (0.4 km) from the shatter ring watch for a pair of holes on the right. The holes are about 2 feet (0.6 m) deep. They formed when a thin shell of solidified lava fell into gaseous pockets concentrated above the flowing lava. In active flows, gas pockets like these inflate the thin, solid crust. There are several more holes along the trail. It pays to use caution when walking across a trail-less, vegetated pahoehoe flow like this one. Imagine finding one of these by accident, or worse, a skylight!

The trail climbs up a bank to briefly leave the lava. This is the pre-lava surface. The blocks of pale pinkish-gray rock in the trail tread are typical Mount St. Helens dacite, deposited here by lahars

A thin, inflated shell of pahoehoe basalt collapsed into gaseous blisters in the lava flow after it cooled. The hole at the lower right is about 3 feet (1 m) across, and the shell of lava is only 6 inches (15 cm) thick.

and pyroclastic flows between 20,000 and 13,000 years ago. The rock is speckled with small white plagioclase and needle-shaped hornblende crystals. Dacite melts at a lower temperature than basalt, and rocks like these became the xenomelts in Ape Cave.

After a short off-lava stretch, the trail regains the flow for another 300 yards (275 m) before again climbing a slope above the flow. The trail joins an old road. There are so-so exposures of pre–Cave Basalt lahar and pyroclastic-flow deposits in old roadcuts. After again descending onto the thinly forested basalt, you quickly reach trail's end at two deep pits that formed where the flow surface partially subsided into the tube below. The southern one bottoms out in a small, evil-looking hole, where you'll find the steel ladder climbing out of the north end of Ape Cave. This is a good place to wander around to look at flow textures. There is some fine ropy pahoehoe near these pits.

8

Looking Down the Gun Barrel

JOHNSTON RIDGE

May 18, 1980 was a glorious spring day in the Cascade Range in southwestern Washington. Gray jays flitted among the subalpine firs rising above the dense undergrowth. The winter's snow was largely gone on the south-facing slope of an unnamed 4,300-foot-high ridge, but pockets remained on the forested north side. Elk that had followed the retreating snowline browsed in a clear-cut. To the south, 4.7 miles (7.6 km) away, loomed Mount St. Helens, its snowy flanks discolored over the past six weeks by muddy gray ash from eruptions of steam carrying older volcanic rock. A huge bulge below the summit protruded toward the north. The steaming volcano seemed to be straining to reach across the densely forested valley of the North Fork Toutle River to join the ridge.

By 8:32 a.m. the sun had already been up for three hours, its rays warming the solitary US Geological Survey geologist keeping watch over Mount St. Helens from an observation post called Coldwater II on the ridge crest. His name was David Johnston. He had volunteered to take his colleague Harry Glicken's slot so Harry could get to an appointment. Gerald Martin, a volunteer ham-radio operator with the Washington Department of Emergency Services, sat by his camper on the crest of the next ridge to the north. He had a higher vantage point and could look down to where David Johnston, 1.5 miles (2.4 km) away, continued with his survey of the distended bulge.

In an instant the huge bulge collapsed in a gigantic landslide. At nearly 0.7 cubic mile (2.9 km³), it is perhaps the largest landslide ever witnessed by humans. Gravity pulled the oversteepened north flank of the mountain into Spirit Lake and across the North Fork of the Toutle River, in the process generating a 5.1 magnitude earthquake. A portion of the landslide of gigantic blocks of rock, dust, nearly the entire mass of the Forsyth and Toutle Glaciers, and broken trees charged 1,150 feet (350 m) up the slope toward Coldwater II at an estimated velocity of 70 to 150 miles (113 to 241 km) per hour.

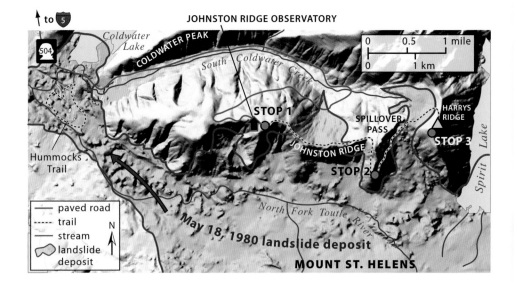

GETTING THERE: The Johnston Ridge Observatory, on the crest of Johnston Ridge, is at the end of Washington 504 in Mount St. Helens National Volcanic Monument. If southbound on I-5, head east at exit 60 (Toledo) on Washington 505. After 16.3 miles (26.2 km) turn left (east) at the junction with Washington 504 and proceed about 40 miles (64 km). If northbound, leave I-5 at exit 49 in Castle Rock and head east on Washington 504 for about 51.5 miles (83 km). Passes, which can be purchased at the observatory, are required to access the observation plaza or to enter the observatory; see the national monument's website for further information. The observatory is more than 4,230 feet (1,290 m) above sea level and closed in the winter. No rock collecting is permitted.

Stop 1 is at the top of the wheelchair-accessible Eruption Trail 201, which begins at the observation plaza at the building entrance. It is 0.2 mile (0.3 km) to the hilltop vista. To reach stops 2 and 3, walk the unpaved Boundary Trail 1, which branches off the Eruption Trail east of stop 1. It is 1.8 miles (2.9 km) to stop 2 at Devils Elbow and 3.8 miles (6 km) to stop 3 atop Harrys Ridge, at 4,800 feet (1,460 m). The elevation gain between stop 1 and 3 is about 500 feet (160 m) and is not steep.

A geologist believed to be Harry Glicken observes the bulge (upper left) on Mount St. Helens from the Coldwater II observation station on May 17, 1980. Glicken would be killed June 3, 1991, in a pyroclastic flow on Japan's Mount Unzen. —Courtesy of US Geological Survey

In places, the landslide overtopped the ridge. The sudden release of confining pressure provided by the bulge's mass uncorked a hot, gas-rich dome of magma that had risen into the volcano's interior. Driven by the expanding gases, the depressurized magma exploded in a violent eruption that headed laterally to the north. This hot pyroclastic surge, consisting of 600°F (320°C) gas, ash, and rocks, blasted outward at an estimated 300 or more miles (480 km) per hour and rapidly overtook the landslide mass.

The surge, consisting mainly of superheated air and a small amount of rock and ash, rose over and crossed the ridge occupied by Coldwater II less than thirty seconds after the eruption began, well ahead of the landslide, and continued northward another 8 miles (13 km), essentially destroying everything in its path. David Johnston yelled into his radio "Vancouver! Vancouver! This is it!" before he was swept away by the stone-laden blast. Martin saw it coming. He reported, "The camper and the car just over to the south of me are covered," referring to Johnston. "It's going to get me, too."

In another few seconds, it did. There was no escape for either of these men; they likely died instantly. They were among the fifty-seven people who lost their lives, most due to ash asphyxiation, in the first few seconds to minutes of the catastrophic eruption.

Johnston Ridge Observatory is near the Coldwater II site. This vignette will not belabor the well-known and oft-told story of May 18, 1980, or of the subsequent eruptions of Mount St. Helens. Consider buying a descriptive book in the observatory's bookstore to bring along on your hike. The observation plaza provides a stunning view into the maw of the fuming volcano and the landslide-filled Toutle River valley; it can serve as an alternate stop 1. The plaza is a fine place to get a unique view of a powerful natural force. Since the plaza can be very crowded, I recommend heading up the paved, gently rising Eruption Trail 201 to the 360° viewpoint on the ridgetop.

From here, an unobstructed view shows the gaping crater left behind when the giant landslide decapitated the volcano and the ensuing eruption reamed out a crater. Domes of pasty dacite lava

Mount St. Helens from the plaza outside Johnston Ridge Observatory. People often mistake blowing dust, picked up by swirling winds inside the crater, for eruption clouds.

have partially refilled the crater during periodic eruptions. The volcano will probably regain its pre-1980 Fuji-like appearance. Some of the growing domes will overflow the crater and cascade down the mountain's slopes as pyroclastic flows. These are hot, dense flows of glowing ash, rock, and gases. You may see small white gas plumes from fumaroles rising off the cooling rock, particularly early or late in the day when the sun is low in the sky. Most of the gas is steam from rain or snowmelt interacting with the cooling lava, but some is from the cooling, shrinking lava itself, principally carbon dioxide, sulfur dioxide, and hydrogen sulfide. The May 18 landslide and pyroclastic surge came directly toward this location, and the slopes around you are veneered with the deposits.

The paved trail loops back to the parking lot in another 0.25 mile (0.4 km), passing the memorial to the victims of the eruption. The route to stop 2 and 3 branches right on the unpaved Boundary Trail 1 about 0.1 mile (0.2 km) beyond the ridge's high point, just past the memorial. It heads out on Johnston Ridge for glorious views and firsthand encounters with deposits related to the eruption. Walk as far as you wish, but a good destination is the summit of Harrys Ridge.

Mount St. Helens consists principally of dacite. Dacitic magma is cooler and more viscous than andesite, and much more so than basalt. Dacite lavas are never runny. When dacite erupts as a lava flow it looks like a creeping pile of hot rock. Dacite more typically erupts as volcanic ash. Gases in the viscous magma are unable to escape the magma easily. As the magma rises into the less pressurized upper level of the crust, huge volumes of gas, dissolved in the magma deep beneath the surface, come out of solution. Gases accumulate as microscopic bubbles, or vesicles, in the magma as it nears the surface in a process called vesiculation. The magma takes on the appearance of foam or froth with only millimeter-thick walls of magma separating the myriad bubbles. When the magma reaches atmospheric pressure at the surface, the gases expand rapidly and shatter the solidified magma around the vesicles. The result is a powerful, explosive eruption that carries large volumes of shattered rock fragments (ash) and larger rock blocks honeycombed with vesicles (pumice).

Many of the pumice clasts along the ridge are angular, just as they were when they exploded out of the vent. Others are rounded, probably due to collisions with other pumice clasts in the eruption plume. Use your hand lens to get a good look at the long, thin tubular bubbles in the pumice. These often make up the bulk of a rock fragment. Pumice can float on water if its vesicles aren't saturated

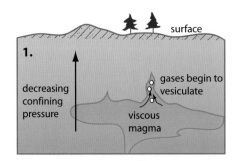

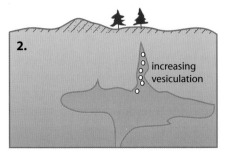

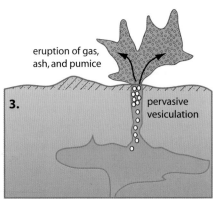

Viscous magma vesiculates as it rises in the crust. The erupted rock, filled with gas bubbles, is pumice.

with water. If you have a container handy and some water to spare, try it. Pick dry clasts from the top of the deposit. The glassy rock on a fresh pumice surface consists almost entirely of stretched-out vesicles, evidence for deformation in the viscous, bread dough–like magma prior to eruption. Black needles are hornblende, a mineral found in almost all dacite.

Views into the crater improve as you go farther east on the moderately level ridge crest. Once forested, all the timber along the ridge crest was swept away by the surge of hot gas peppered with pumice fragments. A scattering of splintered stumps and logs remain, lying as they fell in the scorching hurricane of stone. Branches and bark were abraded, and wood surfaces facing the vent were charred. Below Coldwater Peak, across the valley to the north, bleached logs lie in ranks, all felled in the same direction by the blast. In places a few silver snags still stand, having been protected from the force of the blast by rocky ribs and minor ridges. Though they weren't knocked over, they did not survive the blistering heat. The forest has begun to grow again; a few 10-foot-tall (3 m) alders cast a bit of shade on hot days.

Just beyond a small saddle with a sign that says "mountains can move" you'll encounter hummocks of debris left by the landslide after it crossed the Toutle and charged up the south flank of Johnston Ridge. In places it had sufficient momentum to overtop the ridge and spill down toward South Coldwater Creek. Hummocky topography is characteristic of large landslides; the hummocks were deposited as the landslide velocity slowed. The hummocks along the crest of Johnston Ridge consist largely of blocks of relatively young dacite from the north face of Mount St. Helens. Vegetation has again reasserted itself, layering a tenuous green fuzz over the devastated landscape.

Within seconds of the lateral blast, a towering vertical column of ash and pumice erupted skyward. It reached 12 miles (20 km) into the atmosphere in less than ten minutes and lasted nine hours.

Hummocks left by the giant landslide along the crest of Johnston Ridge.

Trees knocked over in the early morning of May 18, 1980 lie on the south slope of Coldwater Peak.

Pumice, ash, and rock rained out of this eruption cloud, but the wind kept most of it from falling along this portion of the ridge; thick deposits begin just a bit farther east, beyond the bounds of this hike. Pale-tan, 1- to 2-inch-long (2.5 to 5 cm) pumice fragments are found along Johnston Ridge, but these were carried here by the pyroclastic surge. Their presence marks areas that were not veneered by the landslide. You'll really begin to notice them where the trail makes a sharp right and begins a traverse south, directly toward the looming volcano and stop 2.

Stop 2 is the prominent, white, pumice-covered spur called Devils Elbow. The last 0.2 mile (0.3 km) to Devils Elbow is narrow, running across a steep, bare, rocky slope. The rock along this stretch of trail consists mostly of lava and lahar deposits of long-gone volcanoes from millions of years ago. Some of the thin lava flows have prominent vesicles that can be filled with secondary minerals. The lahar deposits are weathered and fairly soft, but their characteristic

Pumice clasts and darker-gray dacite blocks litter the surface of Johnston Ridge. The largest of these is around 4 inches (10 cm) across.

Close view of the stretched-out vesicles and spunglass texture in a pumice clast from Mount St. Helens.

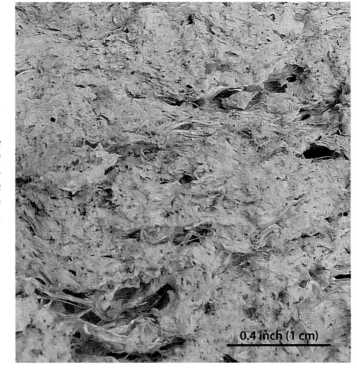

0.4 inch (1 cm)

Huge rafts of trees drift about on the surface of Spirit Lake in this view near stop 3.
—Courtesy of Wendell Beavers

Gases, largely water vapor, rise above the lava dome (center) that formed in Mount St. Helens' crater between 2004 and 2005.

cobbles and boulders are evident. Watch for a few brick-red contacts between pale ash layers and lavas, which baked the underlying ash.

Devils Elbow provides the first glimpse east to Spirit Lake. The cone of Mount Adams rises beyond. Rafts of bleached logs may be drifting on the lake surface. These are the remnants of forest destroyed by the landslide and blast. Spirit Lake was inundated by the landslide and displaced to the north when the debris raised its lakebed 200 feet (60 m). The wave that resulted also uprooted thousands of trees that ended up in the lake.

Devils Elbow is a great spot to look out over the hummocky landslide deposit filling the Toutle River valley. The largest blocks of rock within the hummocks exceed 300 feet (90 m) in diameter. The hummocks protrude through the overlying smoother surface of the Pumice Plain sloping down from the mountain. The plain consists of ash and pumice layers as much as 100 feet (30 m) thick. These deposits formed as incandescent clouds of erupted material spilled down the mountainside on May 18, when the gigantic eruption column grew too dense with ash and pumice and collapsed. Others were added during subsequent dome-building eruptions between 2004 and 2008 (as of this writing). New eruptions could occur at any time. Temperatures in these deposits remained at 780°F (415°C) for weeks after the eruption and were above boiling for a couple of years.

Continue on the Boundary Trail past the junction with the Truman Trail. The trail climbs more steeply through landslide hummocks to Spillover Pass, where a portion of the giant landslide crossed over Johnston Ridge into, and then down, the valley of Coldwater Creek to the north. Just beyond the pass the trail leaves the distinctive hummocky, rocky landslide terrain for one littered with pumice from the pyroclastic surge. The spur trail to Harrys Ridge is about 1 mile (1.6 km) beyond Devils Elbow. Turn sharply right and climb to the crest, stop 3, at an elevation of about 4,700 feet (1,430 m). You are above the west shore of Spirit Lake and can look directly into the crater across 4.8 miles (7.7 km) of uninterrupted space. You will likely feel small and vulnerable, as this was the bull's-eye of the great eruption. The blast stripped everything off this ridge—not only the timber but the soil, too. All the pumice and rocks covering the ground are from the paroxysm of May 18, 1980. The ridge's name commemorates Harry Truman, the obstreperous owner of Spirit Lake Lodge, who refused to leave his property and was killed in the eruption.

to Sumner

					paved road	N
0	0.5	1		2 miles	----- trail	
0	1	2		3 km	——— stream	
					andesite lava	

SUNRISE VISITOR CENTER

SUNRISE RIDGE

Frozen Lake

Sunrise Camp

STOP 1

BURROUGHS MOUNTAIN

White River

410

STOP 2

Burroughs Mountain andesite

PARK ENTRANCE STATION

Glacier Basin

BAKER POINT

STOP 3

410 to Yakima

GOAT ISLAND MOUNTAIN

Emmons Glacier

Fryingpan Creek

123

to Paradise and vignette 10

GETTING THERE: Sunrise is at the end of the Sunrise Park Road in Mount Rainier National Park. From the Seattle-Tacoma megalopolis, head for Sumner on Washington 167 (the Valley Freeway). Turn southeast on Washington 410. About 49 miles (79 km) from Sumner, turn right at the Sunrise Road. This leads to the park's White River Entrance. From Yakima and points east, you can reach Washington 410 via US 12 and Washington 123; turn left at Sunrise Road. The entrance station is about 73 miles (117 km) or 85 miles (137 km) from Yakima depending on which route you choose. The Sunrise Visitor Center is 14 miles (22 km) beyond the entrance station. Sunrise Road is open from June or July until the snow flies. Check in advance. This is a national park, so there is an entry fee.

Sunrise is the jumping off point for many trails. It is the highest point (6,400 feet, 1,950 m) accessible by vehicle in the park. Rock hammers are not allowed in national parks, and it is illegal to collect rocks.

Stop 1 is Emmons Vista, 250 yards (230 m) south of the parking lot. Start on the Sunrise Rim Trail and turn left on the Emmons Vista Trail. For stop 2, return down Sunrise Road for 7.8 miles (12.6 km). Watch for an interpretive sign in a pullout on the east (left) side of the road. For stop 3, continue another 3.6 miles (5.8 km) and park on the west side of the bridge over Fryingpan Creek, at the Wonderland Trail parking lot. Walk 300 feet (90 m) to the other side of the bridge.

9

The Life of an Andesite Stratovolcano

SUNRISE, MOUNT RAINIER NATIONAL PARK

Big volcanoes grow. And then they fall apart. Sometimes they grow back. Large volcanoes such as Mount Rainier usually go through periods of cone growth interspersed with collapse and, in the Cascades, glacial erosion. Contours will change many times over the tens or hundreds of thousands of years that eruptions occur at or near a central vent. Sometimes the change is subtle: a lava flow or two will add a bit of material to the volcano's flanks, and glacial erosion is imperceptible over the human life span. At other times change is catastrophic: a large eruption or the collapse of part of the mountain may profoundly alter the appearance of the cone. Rebuilding occurs when lava eruptions fill scars left by collapses. All the large volcanoes of the Cascade Volcanic Arc show evidence of repeated geomorphic changes. Mount Rainier, a typical large andesite volcano, provides a good example of the results of eruption, erosion, collapse, and rebuilding. Sunrise Visitor Center, on the northeast flank, is a great place to see the geologic evidence for these processes.

Sunrise lies in Yakima Park, an alpine meadow 6.7 miles (10.8 km) northeast of the summit crater of Mount Rainier. Sunrise is a starting point for hikers visiting the high country on the east side of the park. The area is deservedly extremely popular. For Mount Rainier alpine scenery reachable by automobile, only Paradise (vignette 10) can hold a candle to Sunrise. There is more rock 'em sock 'em geology here than you can shake a stick at. Alas, we can only focus on a small part of it.

The dominating cones stretching from Lassen Peak northward through Oregon and Washington to Mounts Garibaldi and Meager in British Columbia are the signature volcanoes of the Cascade Volcanic Arc. Washington's isolated, icy giants are a constant presence on the eastern skyline above the Puget Lowland — when the clouds part. They are all stratovolcanoes, some with obvious, young craters on their summits: the two on Rainier are the best-known examples. Baker sports two craters: Carmelo Crater is hidden beneath 280 feet

(85 m) of ice, and the active Sherman Crater, 1,000 feet (300 m) below the summit on the south flank, is an explosion crater (vignette 22). No lava flows have yet erupted there. Adams has a small ice-filled cinder cone on the summit. Glacier Peak's crater is not very evident and was overtopped by a younger dome-shaped extrusion of pasty lava. And we know the story of Mount St. Helens' famous crater, which formed on May 18, 1980 (vignette 8).

Mount Rainier, surveyed by GPS in 1999, 2008, and 2010 and found to be 14,411 feet (4,392 m) high, is the tallest volcano in the Cascade Volcanic Arc. After Mounts Shasta and Adams, it has the greatest volume of any Cascade volcano: estimates of Rainier's volume range between 34 and 48 cubic miles (140 and 200 km³), versus 108 cubic miles (450 km³) for Shasta, and 48 and 79 cubic miles (200 and 330 km³) for Mount Adams.

These big volcanoes are built of hundreds of lava flows erupted over hundreds to thousands of years. Because lava flows fill topographic lows, successive flows tend to lie alongside or overlap earlier ones. Later lava flows stack on top of earlier ones once depressions are filled. But successive lavas may flow down completely different sides of the volcano so that adjacent flows may not record sequential eruptions, and the ages of adjacent lavas may be thousands of years apart. The andesite and dacite lava flows of the Cascade volcanoes are too viscous to flow far. Because they are largely slow moving and pasty, rather than fast moving and runny like Hawaiian-style basalt lavas, they readily fall apart on steep slopes and descend the mountain's flanks as incandescent pyroclastic flows of rock and ash known as block and ash flows. Andesite and dacite lavas typically do not extend very far beyond their vent, so they stack up higher and higher to eventually form the layered volcanic structure (geologists use the term *edifice*) called a stratovolcano. Magma may find its way to the surface anywhere on the edifice via convenient fractures, so eruptions don't necessarily occur at the summit.

Periods of eruptive growth that make a taller and broader cone are interrupted when the edifice falls apart. Because stratovolcanoes are built of outward-sloping, or dipping, stacks of lava, they are prone to gravity's inexorable pull. This is helped along by the circulation of hot water mixed with sulfur-rich gases rising from magma deep beneath the surface. The resulting sulfuric acid chemically alters solid andesite and dacite to slippery, weak clay and — voila! The stage is set for collapse. The intrusion of a large volume of magma can also swell the volcano's sides outward, as was the case at Mount St. Helens in 1980, oversteepening slopes and further aiding the insidious tug of gravity.

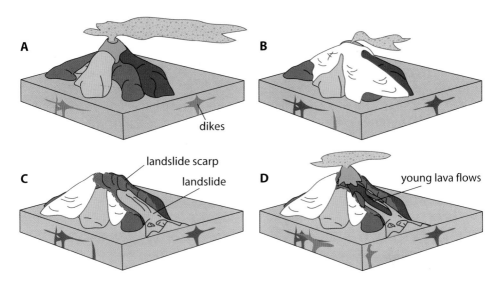

A representative sequence of events in the life of a stratovolcano. Dikes deliver magma to the surface. A stack of lavas grows into a large edifice (A), which becomes tall enough to support glaciers (B). Part of the edifice collapses (C), generating a landslide and lahars in valleys below. (D) Renewed eruptions fill the collapse scar.

In the Holocene period, numerous giant landslides slid off Mount Rainier into the valleys below. These typically generated lahars, or volcanic mudflows, that swept far down the valleys. The most catastrophic event at Rainier was 5,600 years ago, when a small eruption triggered the collapse of about 1 cubic mile (4 km³) of hydrothermally altered rock. A gigantic landslide decapitated the summit and, along with a portion of the northeast flank and the overlying glaciers, roared down the mountainside, leaving behind a horseshoe-shaped landslide scarp. The landslide transformed into a thick lahar called the Osceola Mudflow, which devastated the White and Green River valleys, reaching all the way to Kent, 70 miles (112 km) away. (Some readers may wonder why this lahar was given the proper name "mudflow." The answer: It was named before "lahar" was incorporated into US terminology.) The fast-moving mass covered 212 square miles (550 km²) with mud, rocks, and trees, burying the unfortunate people who surely stood in the way; archeologists found arrowheads under the deposit near Auburn. Muddy, turbulent floods continued beyond the end of the lahar; these reached Elliott Bay in Seattle via the Duwamish River valley, and Commencement Bay in Tacoma.

The scar from the collapse is now partially filled in. After the huge collapse about a dozen eruptions sent lava down Rainier's northeast flank and built two new cones at the top of the gaping landslide scar. Most of that activity has been in the past 2,700 years or so. The summit of Rainier now consists of two nested craters, West and East Craters, consisting of fragmented erupted material, or tephra. These craters rise about 300 feet (90 m) above the plateau marked by the false summits of Point Success (14,158 feet, 4,315 m) and Liberty Cap (14,112 feet, 4,301 m). These two high points define the upper margins of the scarp left by the Osceola collapse. The actual summit of Rainier, Columbia Crest, is the high point on the rim of East Crater, the

US Geological Survey interpretation of the Osceola collapse at Mount Rainier 5,600 years ago. Later eruptions filled in the great landslide scar.

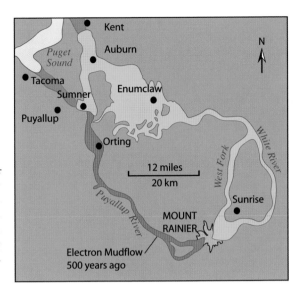

Paths taken by two large lahars from Mount Rainier. Dark blue shows the marine shoreline at the time of the Osceola Mudflow (yellow), about 5,600 years ago. Today a lahar of similar size could extend to Commencement Bay at Tacoma and possibly to Elliott Bay or Lake Washington. (Modified from a map in Driedger and Scott 2008.)

The northeast flank of Mount Rainier seen from stop 1. The red line indicates the approximate scarp of the Osceola landslide; it's dashed where the younger summit craters have filled in the scarp.

youngest of the two craters. The next major geologic event at Rainier could be an eruption — or it could be a collapse of part of the cone.

How high was the volcano prior to the Osceola landslide? The mountain has a somewhat flattened summit plateau, with Columbia Crest above it. In the 1890s, pioneer Rainier geologist I. C. Russell noted the apparent truncation. He posited that the slopes of the outer flanks of the volcano once projected upward to a point in the air well above the modern summit, postulating that the mountain had once been about 2,000 feet (610 m) higher before it was "beheaded," presumably in a paroxysmal eruption. US National Park Service publications picked up Russell's idea, and the notion of a formerly 16,000-foot (4,875 m) Mount Rainier became a part of the mountain's mythology, much repeated today in guidebooks and on the Internet. Don't you believe it!

In the 1960s, other geologists refined Russell's measurements and arrived at a onetime elevation between 15,000 and 15,500 feet (4,575 and 4,725 m). There are significant problems with these contentions of a higher summit, as they do not take into account a more modern understanding of how stratovolcanoes are built. The outward-dipping lava flows on the outermost flanks of the volcano that provided the projections are of different ages, and there is no reason to assume they erupted from the same summit vent, or that they ever would have joined at a common summit. The volcano has not had a simple history of *grow big, fall apart, grow a little more*. We don't know how many collapses have occurred and carried off parts or even possibly the entirety of the summit, nor the location of older vents on the summit. Prior to the Osceola landslide the summit was certainly higher than the scarp left by the landslide, but whether that was 500 or 1,000 or 1,500 feet (150 or 300 or 450 m), we cannot tell. Perhaps Mount Rainier was as high as Russell contended, or even higher, or it may never have exceeded the current elevation. It is misleading to bandy about specific figures that oversimplify the complex history of this great mountain.

Recent fieldwork by US Geological Survey geologists indicates that the most recent magmatic eruptions at Mount Rainier occurred about 1,000 years ago. There were reports of an eruption in 1894. If something actually occurred, it may have been a minor steam eruption that carried older, fine-grained dark rock fragments upward in the plume, as there is no geologic record of it.

At stop 1 there are two stone-walled vistas; go to the second one for the best unobstructed views of the entire northeast side of the mountain. Here, the vast bulk of Mount Rainier seems to dominate the whole world.

Aerial view looking south over the top of Rainier's East Crater; snow blows downwind to the left. West Crater's rim is on the right. In the distance, left to right, are the stratovolcanoes Adams, Hood, and St. Helens. —Courtesy of ©Steph Abegg

The vista is perched on the brink at the top of the 496,000-year-old Burroughs Mountain andesite. Fieldwork indicates this lava probably erupted from a set of dikes to the southwest. This lava is one of the earliest to originate at Mount Rainier, when the volcano was only a fraction of its current size. The lava flowed more than 7 miles (11 km) and banked along the margin of a glacier filling the White River valley. The glacier confined the flow, which ponded to the inordinate thickness of more than 1,150 feet (350 m). Its surface forms the anomalously flat Yakima Park meadow, and it underlies Burroughs Mountain.

The Emmons Glacier covers the northeast flank of Rainier, flowing from the summit into the White River valley. The glacier is the largest in the lower forty-eight states by surface area, covering 4.3 square miles (11 km²). It is named after US Geological Survey geologist Samuel Emmons, who surveyed the mountain in 1870. As of this writing the rubble-covered terminus was at 4,900 feet (1,490 m) of elevation. The rocky debris on the glacier surface is from a series of large landslides that fell off the north face of Little Tahoma Peak, the craggy satellite rising above the glacier's south margin, in December of 1963. About 14 million cubic yards (10.7 million m³) of rock fell 6,200 feet (1,890 m) and traveled up to 4 miles (6.4 km) down the White River valley at around 90 miles (145 km) per hour. The mass of rock helps insulate the glacier from melting; it advanced rapidly in the 1980s and has not receded much since then (see vignette 10).

The Emmons and Winthrop Glaciers occupy the steep scar left by the Osceola landslide, extending about 9,500 feet (2,900 m) vertically from the summit plateau to the valley floor below the Emmons terminus. Columbia Crest is the small, flat surface at the summit. If there has been sufficient snowmelt you can see one of the postcollapse lava flows cropping out as a long, skinny string of rock running down the center of Emmons Glacier. Another window in the ice farther north (right) exposes another young andesite outcrop; both provide rock debris for medial moraines extending down the glaciers beneath the outcrops. The triangle of Steamboat Prow consists of pre-Osceola lava flows. The prow itself may mark the lower margin of the scar, but it likely is a feature that the ancestral Emmons and Winthrop Glaciers eroded before the landslide occurred. Across the valley and 1.5 miles (2.4 km) up the White River from stop 1, Baker Point forms a spur protruding northward from Goat Island Mountain, just below the terminus of the Emmons Glacier. Deposits of the Osceola Mudflow were found there 1,500 feet (460 m) above the valley floor!

Binoculars help reveal the stacks of lava flows exposed in Rainier's cliffs. Remnant snow on ledges may help define them. Good ones are in Russell Cliff, above the west edge of the Winthrop Glacier and north of the summit. Hundreds and hundreds of lava flows compose Rainier's immense edifice. The volcano grew slowly taller as flows piled up wherever the summit vent was at the time of the eruptions. Remember, just because one flow lies directly on another does not mean they are sequential. In the interval between them, tephra may have erupted or subsequent lavas may have flowed in a different direction.

You may never be able to absorb the full grandeur from Emmons Vista. Backtrack down Sunrise Road. Carefully pull over at stop 2 in a pullout on the east side of the road. A wall of nearly horizontal andesite columns, part of the Burroughs Mountain andesite, protrudes from the hillside. There is essentially no shoulder beside the andesite columns, so enjoy this place from across the road beside the interpretive sign. These delicate columns, rarely more than 6 to 8 inches (15 to 20 cm) across, formed as the thick andesite lava flowed along the margin of a huge glacier descending from a young

Lava flows are stacked atop each other in the eroded scarp of Russell Cliff, above the Winthrop Glacier.

Mount Rainier. Certainly some of the ice melted, but the water entered the rubbly surface of the lava flow (stripped by erosion long ago) and quickly solidified the outer rind of the lava (see vignette 5 for more on lava flow structure). Cooling propagated into the lava, which shrank, as any hot solid will do as it cools. The decrease in volume was accommodated by fractures migrating into the interior of the flow at right angles to the cold wall of ice alongside the lava. The width of the resulting columns is inversely proportional to the speed of cooling: thin columns indicate rapid cooling. These thin, horizontal columns indicate that we aren't that far into the interior of the flow, and that the ice was where the valley is today. (See vignette 22 for more on column formation.)

Continue down the road. As you near the valley floor you may note scattered exposures of poorly sorted boulders and smaller rock fragments in a matrix of orange- or pink-tinted muddy sediment. These are remnants of the Osceola Mudflow. Cross over the White

Horizontal andesite columns at stop 2. —Courtesy of the
Washington Department of Geology and Earth Resources

River and continue down the valley to stop 3, an exposure of the
Osceola Mudflow in the 15-foot-high (4.5 m) wooded roadcut on the
north side of the road.

This unimpressive exposure consists of 6- to 36-inch-wide (15 to
90 cm) boulders, mostly andesite, in a fine-grained matrix of clay and
sand. When the weather is dry, the matrix takes on a pinkish-orange
tint. The color is due to the chemical alteration of minerals in andesite lava at the summit of Mount Rainier. If the matrix is damp, you
can easily squeeze it into a cohesive ball in your hand; this reflects
the high clay content, as silt and sand are too coarse to compact. The
boulders were suspended in the matrix during the flow.

It is a crying shame that deposits of the geologic gee-whiz
moment in this valley are so subdued and unremarkable. The lahar
left behind unconsolidated, easily eroded deposits that are easily
penetrated by roots. In this moist climate, forests quickly covered
the lahar.

The Osceola Mudflow deposit at stop 3.

You can squeeze the matrix of the Osceola Mudflow into a lump of clay.

When the summit of Mount Rainier collapsed 5,600 years ago, the gigantic landslide of hydrothermally altered clay, mixed with more-solid andesite rock and the overlying glacial ice, roared down the mountain into the headwaters of the White River. At Baker Point, across the valley from stop 1, the landslide sloshed up the mountainside and fell back into the churning mass flowing down the valley. In the valley the landslide transformed into a slurry of mud and rocks held together in a cohesive mass by the high concentration of clay. This dense slug of sediment, the lahar, raced down the valley like a freight train of soggy concrete, leaving a veneer of debris on the valley walls and floor as the body of the flow roared downstream. At the highway bridge over White River, 0.8 mile (1.3 km) upstream from Fryingpan Creek, the lahar was about 700 feet (210 m) thick, and likely the same at stop 1. The lahar lost material at the valley's base and along its margins, while at the same time it picked up much loose alluvial sediment from the riverbed. A characteristic of lahars is that they can grow larger as they move farther from their source, until they finally lose energy and slow down. At Federation Forest State Park, along Washington 410 about 26 miles (42 km) down the White River from stop 3, the Osceola deposit is still around 440 feet (134 m) thick!

The almost unbelievable mobility of lahars is due to the predominance of water-saturated, slippery clay. It is a mistake to think of lahars as superfloods. They consist of 60 to 80 percent rock and fine sediment by volume, and only 20 to 40 percent water. Lahars can be so dense with sediment that density contrast with rock is slight; large boulders can be suspended in the flow, sinking slowly to the bottom only to bounce up to the surface for a time.

The Osceola Mudflow is the best known of the great Mount Rainier lahars. Several others have descended the White River since. Several have been radiocarbon dated to between 1,350 and 1,700 years ago; the youngest of these made it as far as Kent in the Green River valley. Younger lahars are found in other valleys. Around 1500 CE the Electron Mudflow swept down the Puyallup River, burying the site of Orting with 16 feet (5 m) of mud, boulders, and logs and reaching the outskirts of Puyallup. This lahar probably took only an hour or so to traverse the 38-mile (61 km) distance from the volcano's northwest flank to its farthest reach.

Hot steam and sulfurous volcanic gases percolate upward through Rainier's edifice. Alteration continues to change solid andesite lava and tephra to clay. It is probably inevitable that Mount Rainier will again collapse and send a large lahar into the growing population centers in the valleys at its feet. Which valley will be inundated by the next lahar? When? Will we learn to prepare for one of these low frequency but high consequence volcanic events as urbanization continues to spread ever closer to the great volcano? These questions remain to be answered.

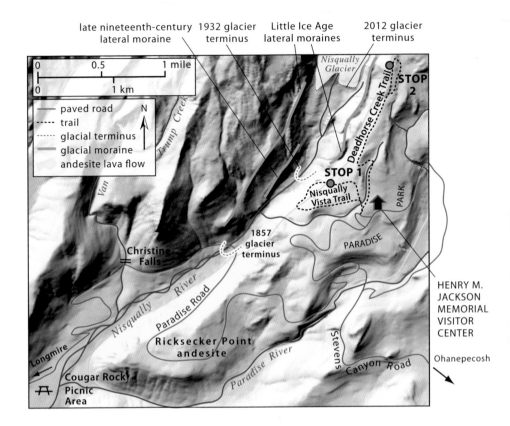

late nineteenth-century 1932 glacier   Little Ice Age   2012 glacier
      lateral moraine   terminus   lateral moraines   terminus

Nisqually Glacier

STOP 2

Deadhorse Creek Trail

STOP 1

Nisqually Vista Trail

1857 glacier terminus

PARADISE

PARK

Trump Creek

Van

Christine Falls

Nisqually River

Paradise Road

Rocksecker Point andesite

Paradise River

Stevens Canyon Road

Longmire

Cougar Rock
Picnic Area

HENRY M. JACKSON MEMORIAL VISITOR CENTER

Ohanepecosh

0   0.5   1 mile
0   1 km

—— paved road   N
----- trail
····· glacial terminus
—— glacial moraine
   andesite lava flow

GETTING THERE: Close views of Nisqually Glacier can be reached from the Henry M. Jackson Memorial Visitor Center in Mount Rainier National Park. From the west side of the Cascades, proceed south on Washington 7 (exit 133 on I-5) from Tacoma, turning left (west) onto Washington 706 at Elbe on Alder Lake. This route becomes Paradise Road. The visitor center is 73 miles (117 km) from Tacoma. From the east, take US 12 from Yakima, turning right (north) on Washington 123 and then left on Stevens Canyon Road, which merges with Paradise Road, for a total distance of about 90 miles (145 km). Alternatively, 17 miles outside of Yakima continue west on Washington 410 from US 12. Turn left (south) on Washington 123 and right on Stevens Canyon Road, which merges with Paradise Road. This trip is about 100 miles (160 km).

From the visitor center, the Nisqually Vista Trail goes to stop 1 and is the easier to reach of the two stops. The trailhead is at the Paradise lower parking lot, the large, looped

10

Debris Flows and Disappearing Ice

NISQUALLY GLACIER

What will happen to Mount Rainier's signature glaciers as global climate warms? Will they disappear entirely? What downstream effects can be expected if recession continues or accelerates? Nisqually Glacier can serve as a case history since it is similar to Rainier's other glaciers. The discussion is pertinent whether you visit both stops or only one, though hiking to both stops in one day is reasonable. Snow lingers late in Paradise. An August visit, when the meadows are alive with flowers, is usually best. Stop 1 at Nisqually Vista will melt out before the higher-elevation stop 2 at Glacier Vista. The hike to stop 2 must be among the very easiest ways to get close to glacial ice in the Lower 48, and it is well situated for additional rambling higher on the mountain and for return loops via the Alta Vista and Skyline Trails.

Of all the ice-covered area in the Lower 48, Mount Rainier constitutes 25 percent—35 square miles (91 km²). The volume of ice on the volcano exceeds that of all the other stratovolcanoes in the Cascade Range combined. There are thirty-three named glaciers and numerous other bodies of year-round ice on Rainier.

The last Pleistocene glaciation was the Fraser glaciation (see vignette 11), which included a period of extensive glacial advances out of the Cascades, as well as the Vashon glacial advance in the Puget Lowland. Between 30,300 and 20,800 years ago, alpine glaciers

parking lot just west of the visitor center. The paved trail is 1.4 miles (2.3 km) round-trip. There are four viewpoints with guardrails; pass the first one for better views at the others.

The trail to stop 2, at Glacier Vista, is longer, but it gets you much closer to the glacier. The Deadhorse Creek Trail begins at the same trailhead and is paved (but with rough spots and steps) until it joins the Skyline Trail (turn to the left), which, after 650 feet (200 m) or so, leads to the spur trail to Glacier Vista. Follow that for about 600 feet (180 m). The round-trip hike is about 3 miles (5 km), and the elevation gain is roughly 940 feet (290 m). You cannot get a view down onto any of the glaciers at Paradise without taking a trail. There is a profusion of them in the area, so pick up a trail map at the visitor center.

In 2013 the terminus of Nisqually Glacier was 0.8 mile (1.3 km) away from Nisqually Vista.

The glacier seen from Nisqually Vista in 1970. —Photo by Rocky Crandell, US Geological Survey

extended far down their valleys toward the Puget Lowland. At its maximum extent 22,000 years ago, the Nisqually Glacier reached Elbe, 23 miles (37 km) from stop 1. These alpine glaciers had begun to retreat before the Cordilleran Ice Sheet's Puget Lobe crept down from the north and covered the Puget Lowland with Canadian ice between 19,400 and 16,400 years ago.

During the Holocene there have been several cycles of glacial advance and retreat at Mount Rainier. Rising global temperatures at the end of the Pleistocene led to a marked decrease in glacial extent; by about 8,500 years ago glaciers were probably a little smaller than they are today. A series of small regional advances over the past 7,000 years is referred to as the Holocene neoglaciation. Dated moraine deposits show that advances separated by

The west margin of the Nisqually Glacier just below the confluence with the Wilson Glacier in 1951, seen from the spur trail to Glacier Vista. The comparison of photos taken since 1890 from the same position indicate that the ice thinned between 1890 and 1951 and then thickened until 1965, but not to 1890 levels. In 2012 the ice thickness was back to 1951 levels and was predicted to steadily thin as the climate warms. Compare this photo with the one on page 162.

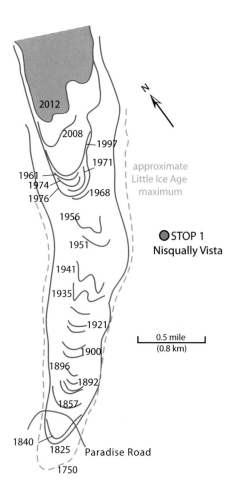

Recession of the Nisqually Glacier, with selected years noted. There were small advances and retreats from the 1950s to the 1970s, but since 1976 recession has been continuous. By 1997 the terminus had melted beyond the 1961 position. (Compiled from aerial photos and National Park Service data.)

warming intervals occurred 4,600 years ago, and between 2,600 and 2,300 years ago. The Little Ice Age, the most recent, beginning in the mid-fifteenth century, was a time of modest cooling over most of the northern hemisphere. Most glaciers thickened substantially during this ice age and advanced up to 2 miles (3 km) beyond early twenty-first-century termini. After this latest advance ended around 1900, Cascades glaciers retreated, though some advanced slightly in the 1960s and '70s. Studies at Mount Rainier show that the glaciated area over the past century has decreased by 27 percent, with a 25 percent loss of volume, and glacial thinning and retreat continue. Loss of ice volume is shown by thinning glaciers and is the critical factor demonstrating loss of ice. It is occurring at a faster pace than terminus retreat. Elsewhere in the Cascades, of the forty-seven glaciers monitored by the North Cascades Glacier Climate Project, six have disappeared since 1995. By 1992 termini at all 107 Cascades glaciers monitored by all researchers were in retreat, and this trend continues. Recession is easily detected in comparison photographs taken only a few years apart.

The Nisqually is probably the most-studied glacier in the United States. It has been surveyed intermittently since the early 1900s, and since 1931 US Geological Survey scientists have photographed it from established stations at approximately five-year intervals. Changes in ice thickness are readily apparent by comparing photos. Since 1900 the terminus has retreated 1.2 miles (2 km). The glacier lost about 15 percent of its area between 1913 and 2009; 30 percent of that loss occurred between 1985 and 2009. At present the terminus is receding 3.3 feet (1 m) per year, the fastest rate ever measured, and the glacier has reached a historic minimum extent.

Not all is gloom and doom. Due to Mount Rainier's great height, it is highly unlikely that the Nisqually Glacier will disappear entirely. However, that may provide only marginal comfort for people who like to visit glaciers and see salmon spawning in the Nisqually River, and the people of Tacoma and other downstream municipalities who depend on a minimum volume of glacier ice to supply water for domestic and industrial use, and for hydropower.

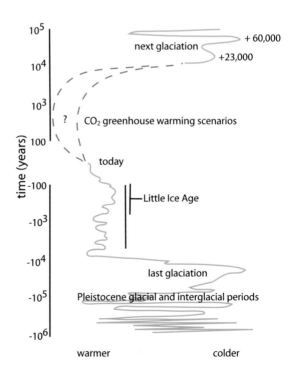

Past and future temperature trends. The lower part of the curve (prior to 10,000 years ago) is based on paleoclimatic evidence from deep-sea cores and glacial evidence. The Holocene part of the curve (the last 10,000 years) is based on variations in mountain glaciers and other types of climate data and historical records. Projections for the next glaciation are based on Milankovitch (Earth orbital) variations. Dashed curves (red) show two of several scenarios for future warming based on assumptions about human-generated greenhouse gases in the atmosphere. (Graph by S. C. Porter, University of Washington.)

Rapid glacial recession has direct geologic and economic effects at Mount Rainier National Park. Annual temperatures have risen 1.5°F (0.8°C) in the Pacific Northwest since 1920. Temperatures are projected to rise an additional 3.6°F to 7.2°F (2°C to 4°C) or more through the twenty-first century. Climate change is also projected to cause wetter winters and drier summers. Most models predict slight increases in yearly precipitation. In addition to the diminishment of the park's iconic glaciers, loss of glacier ice exposes large volumes of glacial debris to erosion. The steep inside slopes of lateral moraines, formerly buttressed by glacial ice, wash downslope in heavy rain or rain-on-snow melting events, choking rivers with sediment of all sizes, including large boulders. In extreme cases, sediment may concentrate in fast-moving wet slurries of rock and finer sediment. Called debris flows, these more closely resemble fast-moving wet concrete than floodwater. They are so dense with mud that they can float large boulders. Debris flows can have great destructive power. They have a nasty ability to bulk up with loose river sediment and become even larger as they move downstream.

Downstream of Rainier's glaciers, reduced channel gradients below canyons and at tributary junctions slow the rivers and streams, reducing their carrying capacity and causing sediment to settle out of the water onto streambeds, raising their levels. Called aggradation, this process occurs in normal floods but is accelerated during debris flows. The normal rate of aggradation in the park's rivers is around 1 foot (0.3 m) or so each decade. A single debris flow in 2006 deposited 6 feet (2 m) of sand, gravel, and boulders in the Nisqually's channel, the equivalent to sixty years worth of gradual sedimentation. The bed of the Nisqually River is now 29 feet (9 m) higher than the administrative and tourist center at Longmire, on the north shore of the river channel. Aggradation encourages the river to jump its banks and flood the valley margins at lower elevations. Concrete-reinforced berms were constructed to keep water out of the historic community, which may need to be relocated.

The 2006 debris flows and floods were spawned when 18 inches (46 cm) of rain fell in a day and a half. Roads, trails, campgrounds, bridges, and utilities were destroyed throughout the park. Damages exceeded $36 million, and the park remained closed for six months. Rapid glacial recession and collapsing moraine walls exacerbated the destruction, loading floodwaters with highly erosive and dense masses of rock and sand, and thousands of toppled trees become floating battering rams. The effects are not limited to the park. Rivers gradually move sediment downstream, away from the mountain and into settled lowland valleys. River valleys become choked with the debris, which only exacerbates flood hazards in settled areas.

Stagnating glacial termini in low-gradient valleys can trap large volumes of meltwater in cavities in and beneath the ice. During heavy rainstorms, or when a summer heat wave adds a sudden influx of meltwater, these "lakes" can release glacial outburst floods, also called *jökulhlaup*, Icelandic for "glacier burst." These outburst floods quickly bulk up with riverbed sediment to become debris flows, sometimes undermining the walls of moraines to incorporate even more sediment. Many outburst floods have occurred at Mount Rainier, including in the Nisqually drainage. The resulting debris flows damaged the old Paradise Road bridge four times between 1926 and 1955. The bridge crossed the river downstream of the glacier. An outburst flood finally destroyed it in 1955. The new bridge, with a clearance of 80 feet (24 m), has so far, survived several outburst floods.

The debris-choked and severely aggraded Nisqually River channel at Longmire after the 2006 floods. The road across the river is washed out.

In October 1947, a rainstorm triggered the collapse of the stagnated lower 1 mile (1.6 km) of the Kautz Glacier, releasing ponded water. The resulting debris flow raced down Kautz Creek and buried the road in the Nisqually River valley with nearly 30 feet (10 m) of rocks, uprooted trees, and mud. In ten hours, several debris flows deposited 52 million cubic yards (40 million m³) of debris. In 2001 a much smaller debris flow occurred in warm but clear weather. This one swept down Van Trump Creek, over Christine Falls at the Paradise Road bridge, and right past the Cougar Rock picnic area and campground. That flow left a 4-mile (6.4 km) trail of debris. Another, caused by heavy rain, poured over Christine Falls in 2005, much to the delight of spectators on the highway bridge. It was captured on video and posted to YouTube.

Mount Rainier glaciers will continue to recede with global warming. This will remove more ice buttressing the inner surfaces of Little Ice Age lateral moraines. Glacier termini will continue to stagnate and may trap meltwater. The frequency of debris flows can be expected to increase, as will channel aggradation. Warning sirens are in place at Longmire, Cougar Rock campground, and the Nisqually Entrance to alert visitors to the possibility of outburst floods, debris flows, or an eruption. If you hear one, head uphill, away from rivers and streams.

Some volcano researchers suggest there is a correlation between deglaciation and increased eruptions. This could be due to the decompression of the flanks of volcanoes as ice is rapidly removed during warming periods. If this is a real correlation, the lag time between deglaciation and the onset of eruptions is poorly constrained at Mount Rainier. There was an increase in eruptive activity between 7,400 and 5,000 years ago, 13,400 years after the end of the last glacial maximum at the volcano and 600 to 2,500 years after the glacial minimum early in the Holocene. Elsewhere in the Cascades, periods of glacial loss over the past 2,300 years may correlate better with increased eruptions. Much more worldwide research is needed before this relationship is convincingly established.

On the way to stop 1 you will pass a large block of andesite lava with a large cavity. The hole is an enormous gas pocket. A lahar carried the rock here; the lahar's deposit surfaces the meadow you pass through. The lahar occurred around 5,600 years ago, at or near the same time as the Osceola Mudflow (vignette 9). It began as a small eruption and collapse of the southern portion of the volcano's summit. The landslide rapidly transformed into a lahar as it surged down the Nisqually Glacier. The lahar was huge; its deposit is 800 feet (240 m) thick in the Nisqually River valley at the Paradise

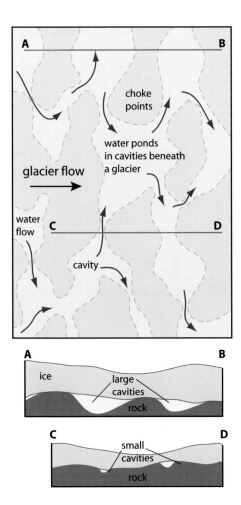

Map view (top) and related cross sections (bottom) showing water flow and storage in glacially scoured terrain. Increases in water pressure due to a sudden rise in water volume melts new pathways or lifts thin, stagnant ice and releases stored water suddenly. (From Walder and Driedger 1994.)

Road bridge and 230 feet (70 m) at Longmire. Deposits are found 24 miles (38 km) away at the community of National, west of the park entrance at Ashford.

Just beyond the andesite block, go right at the fork to reach four viewpoints along the trail directly above the glacially eroded valley of the Nisqually River (the last viewpoint gives the best view). The elevation is 5,400 feet (1,646 m). The huge volcano rises more than 9,000 vertical feet (2,740 m) in the 4.5 miles (7.2 km) between stop 1 and the summit crater. The viewpoints are perched on the rim of the Ricksecker Point andesite, a Mount Rainier lava flow that is only 40,000 years old. It is the largest young lava erupted from the volcano. This lava has glassy margins and slender columns on its east and west sides, suggesting it banked against glaciers in the

Nisqually River and Paradise River valleys (see vignette 22) and cooled relatively quickly.

The Nisqually River flows out of an ice tunnel at the base of the glacier, 0.8 mile (1.3 km) up the valley from the viewpoints. In the mid 1940s the glacier's terminus lay directly below the vista. The 100-foot-tall (30 m) terminus is covered with debris. You may need binoculars to convince yourself that the dark material is actually ice. The thickest ice, measured with ice-penetrating radar, is about 400 feet (120 m) thick 0.6 mile (1 km) up the glacier from the terminus. The Nisqually River issuing directly from the ice is also a good clue that the valley above is covered in ice, the source of the river. The Little Ice Age moraines rise up to 600 feet (180 m) above the milky river on both sides of the valley. Ice no longer supports these steep walls of loose glacial debris, so they are subject to sudden, rapid erosion and even collapse during heavy rainstorms. You may be able to spot deep gullies eroded into their walls.

The view of the Nisqually and Wilson Glaciers from near Glacier Vista. The yellow box is the same area shown in the 1951 black-and-white photo on page 155. The yellow circles are for comparison purposes. —Courtesy of David Midkiff

The entire hike to stop 2 is on the Ricksecker Point andesite, largely covered by orange- or yellow-stained rocks and mud deposited by the lahar. The lahar deposit is not more than 5 feet (1.5 m) thick along the ridge crest to the north, where it slopped eastward over the divide separating the Nisqually River and Paradise River valleys. Along the Deadhorse Creek Trail, watch for trail cuts that expose multiple layers of volcanic ash from Mount Rainier and Mount St. Helens. These are younger than the lahar.

At Glacier Vista (stop 2), elevation 6,340 feet (1,930 m), the rubble-covered terminus of the Nisqually Glacier is below you; it may not be very recognizable as a glacier. Note the concave surface of the ice in the lower part of the glacier, a sure indication that the glacier is retreating. One of the sharp-crested, nonvegetated Little Ice Age lateral moraines is below you. In the mid-1800s the glacier's margin was at that level, and ice rose even higher toward the center. The moraine can be reached via the 0.5-mile-long (0.8 km) Moraine Trail, which you passed on the way here. The same young moraine on the opposite valley wall is very prominent. Mount Adams and Mount St. Helens, 42 and 45 miles (67 and 73 km) away respectively, may be visible to the south.

From here you can continue beyond the interpretive signs to rejoin the Skyline Trail for the return trip, or you can go even higher on the Skyline Trail for ever-expanding views.

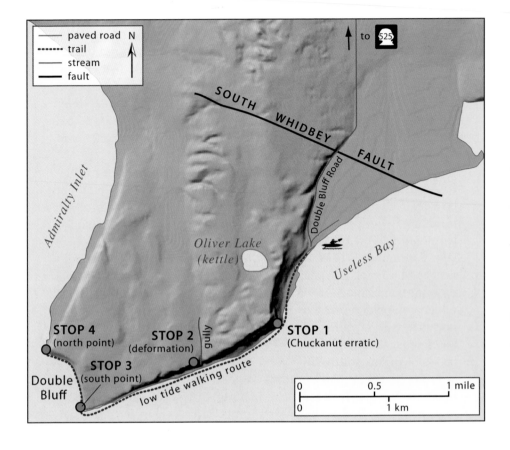

GETTING THERE: Double Bluff is near Whidbey Island's south end. From I-5, either head west on Washington 20 from Burlington or take the ferry from Mukilteo. The Double Bluff Road is about 8.5 miles (13.7 km) west of Clinton on Washington 525 or 19.4 miles (31.2 km) south of Coupeville on Washington 20/525. Turn south off Washington 525 onto Double Bluff Road. It is about 2 miles (3 km) to the beach at Double Bluff State Park, on the shore of Useless Bay. All four stops are along the western end of the beach, away from the parking lot; it's 5 miles (8 km) round-trip to visit them all. As of this writing, the map at the kiosk west of the parking area labeled Double Bluff Beach Access does not accurately show the distance to Double Bluff; it is far beyond the houses, which are along the beach near the sign.

Consult a tide table before trying to walk very far; choose an ebbing tide and allow yourself plenty of time to return to the parking area. The only way to escape a rising tide above 10 feet (3 m) is by scrambling up the sandy bluff and waiting for the ebb. The beach is less than 50 feet (15 m) wide with an 8.5-foot (2.6 m) tide, less if it is stormy, and it is difficult to see the full height of the bluffs when you are that close. A 4.5-foot (1.4 m) tide or lower will allow you to be farther from the bluff so you can see the top. The relevant reference point in the tide tables is Bush Point, on the west side of Whidbey. Binoculars are handy for looking at details in the bluffs.

11

WHIDBEY ISLAND

The huge Cordilleran Ice Sheet, formed from the merger of alpine glaciers in the interior of British Columbia, advanced southward at least six times during the Pleistocene. A large tongue of ice called the Puget Lobe entered the Puget Lowland during each of these glaciations. Successive advances of the Puget Lobe eroded deposits of previous advances, buried them under new deposits, or both. Between the glaciations, during what are called interglacial periods, long intervals of warmer climate prevailed and vegetation and soils covered glacial deposits. Sediment was laid down as dunes and in lakes, streams, and marshes. Fossils and pollen in the sediment came from forests, meadows, and swamps. Steep, eroding bluffs above the sandy beach at Double Bluff contain a record of several glacial and interglacial periods, but you need to walk to the far west end of the beach to see the whole story.

First, let's clarify some terminology. Drift is sediment deposited either directly from ice or from meltwater issuing from ice. Drift includes till, a poorly sorted mix of clay and rock fragments melting directly out of glacial ice without any intervening water transport; outwash gravel and sand, deposited by meltwater streams flowing outward from the glacier; and glaciomarine drift, which is similar in appearance to till, except that the sediment rained down from floating ice to the floor of a sea or lake, so it may be crudely stratified or sorted in places, whereas till is not sorted at all. To be recognized unequivocally as glaciomarine drift, fossil shells of clams or other marine species need to be present in growth position, meaning they are intact and were not moved into place from some other deposit. Shells are often scarce and may require diligent searching.

Because so much water was sequestered in the great ice sheets during glacial maxima, sea level was as much as 400 to 460 feet (120 to 140 m) lower than today. Beneath the glaciers, the mass of ice

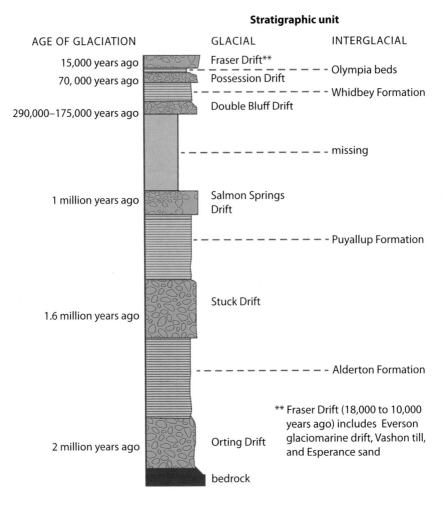

Stratigraphic unit

AGE OF GLACIATION	GLACIAL	INTERGLACIAL
15,000 years ago	Fraser Drift**	Olympia beds
70, 000 years ago	Possession Drift	Whidbey Formation
290,000–175,000 years ago	Double Bluff Drift	
		missing
1 million years ago	Salmon Springs Drift	
		Puyallup Formation
1.6 million years ago	Stuck Drift	
		Alderton Formation
2 million years ago	Orting Drift	
	bedrock	

** Fraser Drift (18,000 to 10,000 years ago) includes Everson glaciomarine drift, Vashon till, and Esperance sand

Idealized Pleistocene stratigraphy in the Puget Lowland. —Courtesy of D. J. Easterbrook

depressed the crust into the mantle. The crust rebounded upward during interglacial periods. The interplay of sea level changes (eustasy) and crustal depression (isostasy) is difficult to unravel in the Puget Lowland. Sea levels have risen due to deglaciation, but rebound has outpaced the rise. As a result, sea level in the lowland and southern British Columbia is actually below some glaciomarine deposits from the times of maximum glacial advance and maximum crustal depression. On Whidbey Island, the highest glaciomarine deposits from the Fraser glaciation are found around 115 feet (35 m) above sea level; they are well exposed in the bluffs at Double Bluff State Park.

The oldest deposit in these coastal bluffs records the Double Bluff glaciation, which occurred between 290,000 and 175,000 years ago. The Whidbey Formation was deposited next in an interglacial period, a time of peat bogs, grasslands, and forests. Mammoths lived here then and left fossil bones, tusks, and teeth. These can be found along the beach today. At Double Bluff, deposits of the next glaciation, the Possession, were eroded away during the subsequent interglacial period called the Olympia, lasting from 70,000 to around 18,000 years ago in the Puget Lowland. It was ended by the last glaciation, the Fraser. Postglacial sediment and soil covers the Fraser deposits.

The Fraser glaciation began about 25,000 years ago when climate cooling allowed alpine glaciers to begin expanding in British Columbia and Washington. However, the alpine glaciers in the Cascades never made it out of the mountains and into the lowlands; they had receded by the time ice from the north arrived. The huge Puget Lobe of the Fraser glaciation reached the Double Bluff area about 18,200 years ago, leaving behind the Fraser Drift (after the type locality on Vashon Island), composed of a number of subunits. The oldest of these is the Lawton Clay, fine sediment deposited in lakes that were impounded in the Puget Lowland when the Puget Lobe plugged the drainage flowing north into the Strait of Juan de Fuca. The slightly younger Esperance sand is sand and gravel from streams flowing out of the advancing ice. Above that is a thin layer of Vashon till, deposited from the base of the ice, and then the Everson glaciomarine drift. As the Juan de Fuca Lobe thinned, seawater flooded into the nascent Salish Sea, floating and disintegrating the Puget Lobe into rafts of ice. The Everson glaciomarine drift fell out of this melting ice onto the seafloor.

Walk 0.5 mile (0.8 km) southwest along the sandy beach, with its sandbars, ripple marks, and shells. At high tide the beach is narrow, but at low ebb it may be 450 feet (140 m) wide. Stop 1 is at a prominent 10-foot-high (3 m) sandstone erratic from the Chuckanut Formation. This rock was carried south at least 28 miles (45 km) from the nearest Chuckanut exposures in the hills south of Sedro-Woolley. From its size we know that ice carried it. It was most likely left behind around 13,000 years ago as the last of the ice melted. Look for a 4-inch-wide (10 cm) palm frond fossil halfway up the west side of the boulder. (See vignette 12 for more about erratics, and vignette 21 for more about Chuckanut fossils.)

The lower half of the 365-foot-tall (110 m) bluff is interbedded sand, silt, and lake clay, probably from the Olympia interglacial period. Streams deposited much of the Olympia. It contains thin,

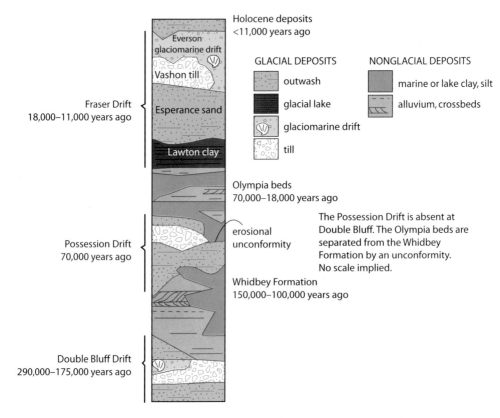

Composite Pleistocene stratigraphy for Whidbey Island. Erosion may have re-moved all or parts of older sediments. An unconformity separates deposits of the Olympia interglacial period and the Whidbey Formation because the Possession Drift had been eroded before the Olympia interglacial. Ages are approximate. (Adapted from Hagstrum et al. 2002.)

sandy strata with crossbeds. Crossbeds are sequential layers of (usually) similar-looking sediment—often sandy—of the same age that lie at angles to each other. Streams generally produce crossbeds by eroding an uneven surface in the sediment of the streambed. An upper bed of sediment is later deposited on that surface. The result is stacked layers that lie at angles to each other, reflecting variations in the stream's energy—high to erode, lower to deposit. Individual beds of sand or clay thicken and thin because streams meander back and forth across the landscape, depositing fine sediment, silt, or sand on the inside of their meander bends, where the velocity is lowest. On the outside of the bends, where the current is greater, finer-grained sediment is eroded.

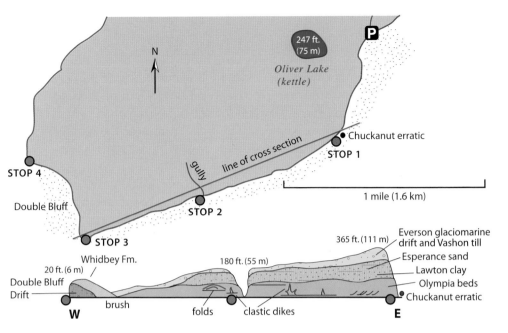

Schematic cross section along the bluffs. Not to scale. Bluff height declines from east to west. (Adapted from Polenz et al. 2006, based on detailed stratigraphy by Stoffel 1980.

The stratigraphy at stop 1 records the most recent transition from an interglacial (Olympia) to a glacial (Fraser) environment.

Some argue that the lower half of the bluffs belong to the Whidbey Formation rather than the Olympia; however, there is no profound erosional unconformity to suggest 82,000 years of missing time (the difference in age between Whidbey and Lawton, which undoubtedly lies above). Radiocarbon dates from near the base of the bluffs are at least 40,000 years old, well within the 70,000-to-18,000-year age range for the Olympia.

The Olympia interglacial period persisted for 52,000 years until the Puget Lobe again buried the area under ice. The advancing ice blocked north-flowing surface drainages and ponded a large lake, Glacial Lake Russell, against the lobe's terminus. Fine sediment deposited on the floor of the lake is the first deposit associated with the Fraser glaciation in the lowlands. This unit is the Lawton clay, overlying the Olympia. At Useless Bay this clay is about 50 feet (15 m) thick. Most of the rest of the bluff is Esperance sand, outwash characterized by layers of sand with a few pebbles, coarsening upward in the bluff and becoming gravelly. Energetic outwash streams spawned by the Puget Lobe deposited the gravel as the lobe entered the lowland and displaced the northern edge of Glacial Lake Russell. A lighter-colored, tan or yellowish deposit in the top 10 to 20 feet (3 to 6 m) of the bluff has a sharp contact with the Esperance sand and may cut into it in places. Binoculars will help here. This is poorly sorted Vashon till, Everson glaciomarine drift, or both.

Continue your hike along the beach, littered with a wonderful variety of erratic boulders, cobbles, and pebbles. The sedimentary layers in the bluffs were at one time much more extensive and likely continuous with similar deposits on Puget Sound's western shores. The sea is steadily eating away at the bluffs, carrying off the small clay, silt, and sand grains so that larger stones are all that remain. Some estimates posit that these bluffs have receded northward 100 feet (30 m) over the past forty years. Active erosion and retreat is evident in the steepness of the bluffs, the many slumps along their length, and the lack of vegetation—erosion is too rapid for trees to get a toehold.

A few fossilized mammoth bones and tusks have been found on the beach, weathered out of the Olympia beds, so keep your eyes peeled! This is state-owned land, though, so the removal of vertebrate fossils for personal possession is illegal. If you are lucky enough to see a fossil, leave it undisturbed or gently bury it and contact the Burke Museum in Seattle or the Western Washington University geology department with, if possible, the GPS location and a photograph of what you found.

Among the highlights of the walk are curious, approximately vertical yellowish-tan clastic dikes within gray sand layers 50 feet (15 m) above the beach. These structures swoop and swirl upward and even cut across each other. You'll start to see them soon after rounding the point at the sandstone erratic, and they continue beyond stop 2. Clastic dikes are analogous to intrusive igneous dikes in that they intruded other geological strata, but they are composed of clasts. These dikes form nearly instantaneously when saturated layers liquefy during ground shaking. When earthquakes shake sediment below ground, grains compact and settle, while water between the grains rises with suspended lighter grains. This muddy water invades overlying layers, leaving intrusions of sediment. The clastic dikes at Double Bluff consist mainly of fine sand.

25-foot-tall (8 m) clastic dikes invade sand west of stop 1.

Geologists with the US Geological Survey suggest the South Whidbey Island fault zone is responsible for the earthquakes. The fault zone, about 5 miles (8 km) wide and composed of several parallel faults, runs from Victoria, British Columbia, through southern Whidbey Island 1 mile (1.6 km) or so north of Double Bluff and continues southwest through Mukilteo at least to Woodinville, and maybe beyond. The largest historic earthquake of this zone was in 1976, when a magnitude 4.7 temblor originated more than 13 miles (21 km) below the surface right here. There have been a dozen shakers since with magnitudes between 2 and 3.8. The fault zone is believed to be capable of generating magnitude 7.5 earthquakes, which could cause major damage in the region. Clastic dikes have been found in research trenches cut across the fault zone in other places.

If clastic dikes reach the surface, they erupt water, mud, sand, and even gravel to form volcano-like mounds up to 2 feet (0.6 m) high. The mounds are called sand blows or sand volcanoes. During the 2001 Nisqually earthquake, many sand blows erupted around the south end of Puget Sound. It would be interesting to find a clastic dike in these bluffs connected to a sand blow. For a sand blow to be preserved, it had to be buried immediately after forming. The eroding bluff surface changes at a rapid rate, and so does the exposure of clastic dikes. A visit to these bluffs every few years increases the chance that you might find a newly exposed sand blow in cross section. You can scramble up the bluffs to get a close look at the dikes in most places. The tan dikes consist mostly of silt, while the layers they invade are gray, thinly bedded sand.

Continue west. Watch for large blocks of clay that have fallen from the Olympia or overlying Lawton deposits. Since clay settles very slowly in water, it indicates deposition in calm water, such as lakes and ponds. Some of the clay blocks are made up of very thin layers, each from a different deposition event. Perhaps these formed when seasonal rains flushed sediment into the bodies of still water.

About 1 mile (1.6 km) from the parking lot, a prominent brush-choked gully reaches the beach. This is a good landmark for stop 2. The bluffs have been steadily getting lower since stop 1 and are only 180 feet (55 m) or so above the beach here. The truly adventurous can scramble up the gully to the top of the bluffs to see a stratigraphic cross section. Starting your ascent just outside the gully's east edge is your best bet. You'll be in a variety of Olympia beds for most of the way, but about 50 feet (15 m) of Lawton and Esperance deposits appear as steep banks of well-sorted clay, silt, or sand with an occasional pebble. These are distinctly replaced at the top by drift composed of pebbles in a fine-grained matrix. This layer may be Everson glaciomarine drift over Vashon till; it is sufficient to say that you are seeing direct evidence of the last glacial ice in the Puget Lowland. It may be difficult to get to the very top of the gully because the drift is slick when wet.

The folds in the clay at the top of Olympia interglacial sediment 100 yards (90 m) west of stop 2 may be related to earthquakes or glaciotectonic stress. The sand and clay layers above and below are not deformed.

About 100 yards (90 m) beyond the gully are two sets of deformation structures in two different Olympia layers. The lower set comprises clastic dikes that invade grayish sand; these are similar structures to the ones we've been seeing since stop 1. The upper set, about 100 feet (30 m) above the beach, is different. These 20-foot-wide (6 m), 20-foot-tall folds are oriented east-west and are clearly layered. The dark-gray to tan beds within the folds are contorted. The folds may be related to earthquakes, but the different morphology suggests they may be due to glaciotectonic stress exerted by the crushing weight of the Puget Lobe of either the Possession or Fraser glaciations. It is possible to climb arduously up loose sand to the base of the folds for a close look.

The Olympia frequently contains thin, sandy crossbeds deposited by streams. These were truncated as the stream eroded its own bed and then quickly deposited new sediment at a slightly different dip. The small holes are insect burrows. The hand hoe at right is 12 inches (30 cm) long.

About 200 yards (180 m) farther west, steep clay beds from the Olympia interglacial period rise from near beach level. Here may be the best opportunity to look closely at the layers left in ever-changing streambeds. Look for places where layers have been truncated by erosion and then covered with slightly younger beds with slightly different dips; these are great examples of crossbeds. They serve as a good illustration that sediment layers deposited in streams and rivers are not necessarily perfectly horizontal, as expected in lake or deep-ocean sedimentary layers.

Farther along, a bed of dark peat appears in Olympia sediment 50 feet (15 m) above the beach. The peat tapers out to the east at the shoreline of the ancient marsh in which it was deposited but quickly thickens to about 3 feet (1 m) to the west. The pure, organic peat is dense and resistant to erosion. It holds moisture in the midst of the vertical sea of sediment, so bushes grow out of it. The vertical surface may be cracked where it has dried out. You may already have found large chunks of dark, flaky peat lying on the beach; it looks like wood but is really made of compressed carbon-rich plants. Most of the black organic matter is composed of brittle, flattened stems of 100,000-to-150,000-year-old horsetails. This peat was buried by more than 100 feet (30 m) of sand and mud and had about 3,000 feet (900 m) of ice slide above it — twice. You'd be flat, too! One 30-foot-long (10 m), 6-foot-wide (2 m) fallen slab of peat has persisted on the beach for many years, lying at an angle to the bluffs well below the high tide line. You might mistake it for an incongruous ledge of rock. Several pieces of wood up to 1 foot (0.3 m) long are embedded in the peat.

After about 200 yards (180 m) or so, the peat pinches out and the bluff lowers. Eventually a layer of bouldery drift comes into view at the top, sloping down toward the west, until it is at eye level. Stop 3 is the 20-foot-tall (6 m) bluff at the southern point of Double Bluff. The layered deposits, the Double Bluff Drift, represent the oldest glaciation preserved in the central and northern Puget Lowland and are the bottom of the stratigraphic sequence in these bluffs. The lowest 2 to 3 feet (0.6 to 0.9 m) is till. There is a distinct contact where it is overlain by shell-bearing glaciomarine drift, though shells are very sparse. Ages obtained from shells indicate the drift was deposited between 178,000 and 111,000 years ago. Near the top of the bluffs there is a sharp contact with overlying Whidbey Formation sand and gravel, deposited after the Double Bluff glaciation had ended. Double Bluff Drift dives out of view somewhere in the brushy interval to the east.

Since stop 1 you have walked past deposits of two glacial advances, the Double Bluff and Fraser. A third, the Possession, has been eroded from this section. Strata from two nonglacial periods, the Whidbey and Olympia, have also been well exposed.

Stop 4 is 0.5 mile (0.8 km) away at the northern point of Double Bluff. Pay careful attention to the tide since you can't escape up these vertical bluffs! The bluffs are primarily Double Bluff Drift; fine sand forms the very base in some spots, probably outwash heralding the advance of the Puget Lobe during the Double Bluff glaciation. The Whidbey Formation lies at the top. Three private stairways descend the bluff. At the stairs farthest north, a reddish-orange layer appears above dark peat about two-thirds of the way up the bluff; a fainter, less-red layer is below the peat. This appears to be oxidized soil, perhaps baked by an underground fire that burned in the peat. The layers can't be reached from the beach, but you may find pieces of the red sediment below the bluff. The pieces I examined were hardened clay and resembled chunks of a somewhat softer-than-ordinary brick, lending credence to the burning peat hypothesis.

The contact (dashed red line) between Double Bluff Drift and Whidbey Formation is visible at stop 3. The inset shows the unsorted composition of the till, containing both cobbles and fine sediment.

Complex stratigraphy greets you at stop 4. Everything in the bluffs is Whidbey Formation: the uneven surface of the Double Bluff has disappeared again, a victim of later erosion. A 15-foot-thick (4.5 m) gravelly layer is sandwiched between sand but tapers to zero thickness around the point to the north. It appears to fill the course of a fairly boisterous stream (only pebbles and cobbles were left behind by the current), which eroded a channel down into river sand. As the channel shifted laterally out of this scene, it buried the gravel with sand, indicating lower stream energy.

If the tide is out far enough, walk out and look back to see how close the houses at the top are to the bluffs' edge. Now think about the estimate of 100 feet (30 m) of recession in forty years. Makes you wonder if the bluffs will last long enough to see the next Puget Lobe cover this place with a new load of drift.

Just east of stop 4 is a layer of black peat with baked red clay above and below.

As a huge stone is sometimes seen to lie

Couched on the bald top of an eminence;

Wonder to all who do the same espy,

By what means it could thither come, and whence;

So that it seems a thing endued with sense:

Like a sea-beast crawled forth, that on a shelf

Of rock or sand reposeth, there to sun itself.

—From "Resolution and Independence,"
by William Wordsworth, 1807

12

GLACIAL ERRATICS OF THE PUGET LOWLAND

Like a badly trained kid, glaciers don't clean up after themselves. Whatever they carry along ends up littering the landscape when the ice melts. Glacial erratics are probably the most startling and visible glacial relics in the Puget Lowland. Erratics are rocks from somewhere else—foreign rocks that do not match the local bedrock on which they lie or that is nearby. *Erratic* is derived from the Latin *errare*, "to wander."

Naturalists have long known about massive, conspicuously out-of-place boulders. It seemed simple to attribute them as evidence for the flood in Genesis. It was only in the 1830s that erratics began to play a significant role in scientific discussion, during the heated debate over whether ice once covered Europe and North America. Charles Lyell, one of the founders of modern geology, described erratics in his groundbreaking *Principles of Geology* (1830), although he ascribed them largely to icebergs floating on water. Charles Darwin, who considered himself first and foremost a geologist greatly influenced by Lyell, wrote at least nine scientific papers on the subject of erratics. He had heard reports of icebergs carrying large boulders far out to sea; thus he deduced that ice rafting was the source of the hundreds of boulders he puzzled over in Tierra del Fuego during the cruise of HMS *Beagle*. (Much later, others showed that these were deposited directly from melting glaciers.)

Any rock fragment that is different from the local bedrock on which it sits is an erratic. If a glacier was the mechanism of transport, we use the term *glacial erratic*. Technically, glacial erratics include the small pebbles and sand grains in glacial till, but these are hardly noticeable to most of us. We tend to fixate on the big ones, the impressive house-sized and larger boulders that are pretty hard to miss. It isn't uncommon for these hulking boulders to be known at least locally by an "imaginative" name, such as Big Rock or House Rock.

Glacial erratics are ubiquitous and easy to spot along the shores of the Salish Sea. These eroded out of glacial till in the bluffs at Point Whitehorn in Whatcom County.

Erratics are useful to geologists. If the provenance, or source, of the rock can be determined, then the direction of ice flow can be estimated. Nothing more easily demonstrates that great ice sheets flowed south out of British Columbia than the presence of hundreds of thousands of Canadian-derived boulders strewn about the Puget Lowland. Some are hundreds of miles from their source outcrops in central British Columbia. Determining an erratic's origin may be fairly simple. For example, quartzite is found nowhere in western Washington or the Cascade Range, though this durable, hard rock is found as pebbles in river valleys throughout central British Columbia, and as pebbles and cobbles in the tills of western Washington. Similarly, conglomerate of the Jackass Mountain Group, found east of Hope, British Columbia, and farther north in the Fraser River valley, composes distinctive erratic boulders in northwest Washington.

It is much more difficult to assess the origin of the salt-and-pepper granodiorite erratics common in the Puget Lowland and slopes of the Cascades and Olympics. These may come from the

British Columbia Coast Mountains, although they could conceivably come from the Cascade Range on both sides of the border. Chemical analyses of these boulders to determine which pluton they came from give equivocal results at best due to the basic similarity of so many of these rock bodies. There is little to no erratic-based evidence that Cascades alpine glaciers ever reached the Puget Lowland, so a Washington Cascades source for these boulders seems unlikely. Erratics also help estimate minimum ice thicknesses; granitic erratics from Canada found at 5,000 feet (1,520 m) of elevation in the Twin Sisters Range, east of Bellingham, show that ice was at that elevation or higher.

Ice sheets advanced out of the British Columbia interior and Coast Mountains into western Washington multiple times. It is

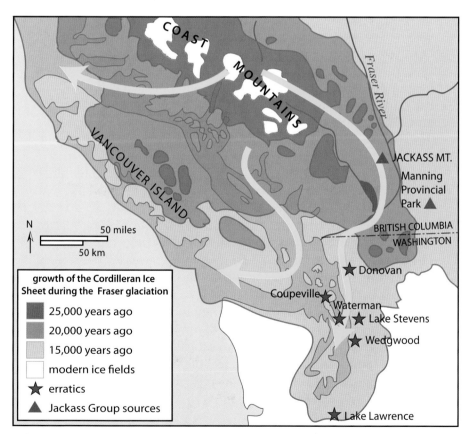

During the Fraser glaciation the Cordilleran Ice Sheet expanded radially from small ice caps covering mountains, such as those found today in British Columbia's Coast Mountains. The Puget Lobe flowed south into the Puget Lowland less than 20,000 years ago and reached its maximum southern extent around 15,000 years ago. (From Booth et al. 2003.)

feasible that any given glacial erratic was picked up and carried more than once by successive glacial advances separated by tens of thousands of years. However, in the interglacial intervals, outwash and other sediments may have buried some erratics, keeping them out of reach of the next glacier to override the area.

Saltwater beaches are perhaps the best and easiest places to look for erratics. If the beach is backed with bluffs of glacial deposits, erratic boulders will be left behind as a lag deposit after storms and tides wash away the finer-grained sediments. Washington's gravel beaches consist almost entirely of pebbles and cobbles transported by glaciers and then worn smooth by glacial, stream, and wave erosion.

Erratics are incorporated into glaciers by a variety of means. Landslides and rocks fall onto the surface of a glacier; this process can deliver really huge rocks, which are then piggybacked for great distances. Overriding ice can scoop up loose surface deposits, such as stream cobbles or older glacial tills. The immense force of flowing ice can pick up, or pluck, large blocks from outcrops, particularly if the rock is already fractured. Rocks carried at the base of a glacier are ground against underlying bedrock and polished in the rasping, gritty, high-pressure environment at the base of a glacier. Boulder surfaces are ground flat by this rough treatment and can be deeply scratched, leaving parallel lines called glacial striations. If the internal flow of ice in the glacier raises a boulder above this destructive zone, it can be carried to a final resting point at the glacial terminus without further damage; some erratics in the Puget Lowland may have traveled 350 miles (560 km) or more.

Not all erratics were deposited directly from a melting glacier. Icebergs can raft boulders along; the boulders drop out of the ice when it melts at sea or in a lake next to a glacier, or the iceberg may be beached on a distant shore. The great Missoula floods, spawned by the catastrophic failure of glacial dams across the Clark Fork River northeast of Spokane, carried rock-laden icebergs across eastern Washington and down the Columbia River (see vignette 2). The rocks dropped from these bergs as the floodwaters ebbed can be found far south of the maximum extent of Pleistocene glaciers. Some small erratics have a completely ice-free origin. Stones entwined in the roots of fallen trees can drift down rivers or wash up on ocean beaches. Once the root-ball rots away, the erratic might really puzzle a geologist. Stones are also carried in the gizzards of birds and the holdfasts of kelp. These, however, are always small and would be pretty inconspicuous. A lahar can carry very large boulders quite a long way down a river valley.

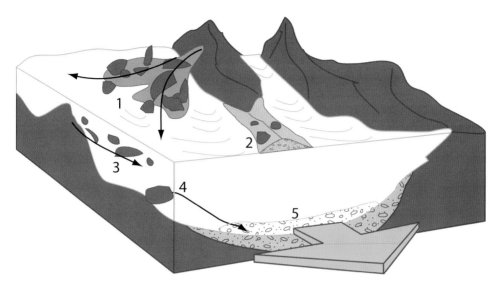

Large rocks are entrained in glacial ice by falling onto the surface in landslides (1), are incorporated from moraines (2), and are plucked off high points overridden by ice (3). Very large rocks are trundled along at the base of the ice (4) and may be ground to small bits. Boulders picked up from river gravel (5) are frozen into the base of the glacier.

Rocks lying on the surface of Black Rapids Glacier are erratics in the making. —Courtesy of John Clague

Larger erratics can be used to calculate the time that has elapsed since the boulder was left behind by ice, which is equivalent to the age of deglaciation in a given region. The chlorine-36 dating method measures the amount of time a rock surface has been exposed to cosmic rays that bombard Earth from outer space. These high-energy particles collide with rock (or any other substance), penetrating 30 feet (10 m) or more into the Earth. Scientists perform chemical analyses to calculate how long a rock surface has been exposed to the rays. A few chips of rock from the uppermost surface provide the material used in this method. It is important that the erratic surface was not shielded from cosmic rays after the glacier left it. This could happen if later sedimentation temporarily buried the rock, or if it was undercut by stream erosion thousands of years later and rolled down a slope to expose a new surface to bombardment. The largest erratics are unlikely to have been moved since they were left by ice. The chlorine-36 method is especially useful for determining the age of continental deglaciation because organic matter for the more familiar carbon-14 dating method is scarce in glacial deposits.

This vignette guides you to a few large glacial erratics scattered around the Puget Lowland, an area which was, at various times, covered by a continental ice sheet as much as 6,000 feet (1,830 m) thick around Bellingham and 3,000 feet (915 m) at the future site of Seattle. The featured boulders are significant either because of their size, their location, or the rock itself. This is by no means a complete list. (See http://nwgeology.wordpress.com/the-fieldtrips/glacial-erratic-field-trips/ for information about other Washington erratics.) Please do not use hammers on these rocks. You may not believe that your visit does much damage, but there is a case history pointing to the damage caused by geotourists. The Sheridan erratic, stranded by a Missoula-flood iceberg southwest of Portland, Oregon, weighed about 160 tons (145 metric tons) when measured in 1950. Visitors have chipped away at the rock bit by bit. Recent measurements estimate that it now weighs 90 tons (80 metric tons). An unbelievable 70 tons (65 metric tons) of rock has been removed by geotourist attrition!

Donovan Erratic, Bellingham

The Donovan erratic is a striking conglomerate rich in rounded pebbles and cobbles. It is from the mid-Cretaceous Jackass Mountain Group in southern British Columbia. The cobbles that characterize the conglomerate were eroded from basaltic and granitic rocks, along with less abundant sedimentary rocks such as argillite and chert, long before the conglomerate was cemented. The ancient rivers that carried these rounded clasts deposited them on a submarine fan in deep water off the continental shelf of North America around 120 million years ago, when the continental coastline was far east of its current position.

Outcrops of Jackass Mountain conglomerate can be easily visited at Manning Provincial Park in British Columbia. A more likely source for the Donovan erratic, however, is north of Hope in the Fraser Canyon near Boston Bar, 100 miles (160 km) distant as the glacier flows. This is a more direct ice route to the Puget Lowland than from Manning. Note that the fractured faces of the Donovan erratic cut right through the cobbles, sectioning them rather than breaking around them. This indicates that the sedimentary conglomerate was thoroughly lithified by deep burial beneath thick piles of sediment and probably experienced a small amount of metamorphism, too.

The Donovan erratic is very hard. However, if you visited an exposure of this same rock in an outcrop, say at Manning Provincial Park, you'd find that you could easily break the weathered surface apart with a hammer. Why such a difference in apparent hardness? Abrasion during transport by the glacier removed soft, weathered exterior surfaces, leaving the hard, unfractured interior. Jackass Mountain conglomerate erratics are distinctive and hard to misidentify; they are commonly found on beaches. Some locals have long maintained that the Donovan erratic is a meteorite. It is

GETTING THERE: Take exit 250 from I-5 in Bellingham. From the ramp turn right (west) on Old Fairhaven Parkway, then right (north) on 30th Street, and right again on Donovan Avenue. The erratic is at the dead end of Donovan across 32nd Street, so park along 32nd and walk up Donovan toward I-5. It is about 0.5 mile (0.8 km) from the I-5 exit to the erratic. Admire the erratic from the street, as most of it sits on private property.

Rounded cobbles are prominent in the conglomerate of the Donovan erratic. The white clasts are well-rounded granodiorite cobbles.

true that meteorites are mistaken for erratics, and in a way, they are the ultimate erratics! The Donovan erratic was once much larger, around 16 feet (5 m) high. It was blasted apart to make way for I-5 in 1965.

Erratics of Jackass Mountain conglomerate are found at least as far south as Federal Way in King County. These rocks are significant because they demonstrate that the Puget Lobe, the last great advance of the Pleistocene ice sheet in this region, came out of the

The 16-foot-high Donovan erratic was dynamited in 1965 to make room for I-5.
—Courtesy of Western Washington University Geology Department

interior of British Columbia past known locations of the Jackass Mountain Group all the way to the Puget Lowland.

The perfectly round, polished, 39-inch (100 cm) sphere of the wonderful "Levitating Sphere" sculpture in front of Kulshan Hall at Whatcom Community College is made from the Donovan erratic. It sits within a 7,000-pound (3,175 kg), water-filled bowl cut from the same conglomerate. The bowl's diameter is less than 0.04 inch (1 mm) larger than the sphere. The 3,000-pound (1,360 kg) sphere "levitates" in the bowl, suspended by a thin film of water, and rotates at the slightest touch. If you visit, be sure to go inside the building to read the interpretive sign on the wall, which has photos of the erratic prior to blasting and an explanation of how the sculpture was made.

The 3,000-pound "Levitating Sphere" sculpture at Whatcom Community College spins with the touch of a finger.

Big Rock, Coupeville, Whidbey Island

There are many glacial erratics studding the landscape of Whidbey Island; virtually the entire island south of Deception Pass State Park consists of glacial deposits. This big erratic in Coupeville is a well-known local landmark. Known as Big Rock, it is about 22 feet (7 m) tall and festooned with ivy. The rock is very green, the product of the low-grade (meaning relatively low-temperature and low-pressure) metamorphism of iron- and-magnesium-rich volcanic rocks, which made it chlorite rich. Such rocks are called greenschist if the minerals were deformed or realigned by metamorphism, forming layering called foliation, but greenstone if not. This erratic is slightly foliated greenschist, and it looks very much like rock exposed to the north, beside the restrooms at the south end of the Deception Pass bridge on Washington 20. Outcrops there, and probably Big Rock, are part of the Fidalgo Complex, a group of rocks that began as a chain of volcanic islands that later were accreted to North America. The greenschist is from the lavas erupted in those islands. If that is where this rock came from, then the Puget Lobe hauled it only 13.5 miles (21.7 km). Chlorine-36 ages obtained from the erratic's upper surface show that it was dropped here around 16,300 years ago. This compares favorably with a radiocarbon age obtained from shells in post-glacial sediment in nearby Penn Cove; they lived approximately 15,700 years ago.

GETTING THERE: From the north, take Washington 20 west from I-5 in Burlington toward the town of Coupeville, on Whidbey Island. From points south, take the Mukilteo ferry to Clinton and follow Washington 525/20 to Coupeville. At Coupeville, turn south on South Main Street at the big pedestrian overpass. The erratic is just behind the shops on the right after 0.1 mile (0.2 km), tucked up against a house and apartments (Big Rock Apartments, of course). This gigantic erratic is on private property but is easily visible from the parking lot.

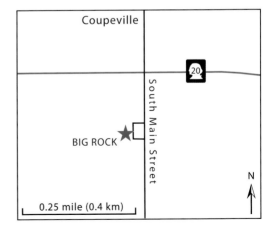

The ivy-covered Coupeville erratic.

The Waterman Erratic, Whidbey Island

The Waterman erratic is slightly foliated greenschist, like Coupeville's Big Rock. The Waterman could be from the Fidalgo Complex exposed at the north end of Whidbey Island. The hulking brute is 38 feet (12 m) high and 60 feet (18 m) long on each side. Its circumference is 155 feet (47 m). After the Lake Stevens erratic (see below), it is the largest known glacial erratic in the state. Its volume is estimated to be 136,000 cubic feet (3,850 m³), and it is estimated to weigh 13,114 tons (11,897 metric tons). For comparison, the Space Needle, including the massive foundation, weighs a mere 9,550 tons (8,664 metric tons). The rock sits in a depression and is well hidden in the forest. It is rounded and quite smooth and has very few cracks. The monolith is vertical on three sides and leans back slightly on the west. There may be a rotting rope hanging down that face, but it's not safe to use.

Chips taken off this rock and dated using the chlorine-36 dating method indicate it melted out of the Puget Lobe 15,500 to 14,700 years ago. This is younger than the Coupeville erratic to the north. Since we know the ice retreated northward, we would expect the Waterman erratic to have been dropped by the ice and exposed to cosmic ray bombardment before the Coupeville erratic; perhaps the Coupeville erratic sat on the glacier surface longer than the Waterman erratic. This discrepancy illustrates that using chlorine-36 dating on erratics may have inherent problems.

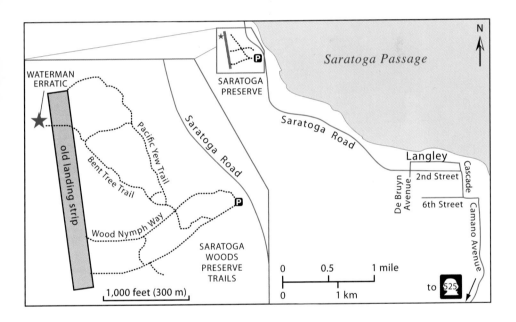

GETTING THERE: See the Coupeville erratic for directions to Whidbey Island. Proceed to Langley via Langley Road or Maxwelton Road off Washington 525 on southern Whidbey Island. Follow Camano Avenue and then Cascade Avenue into town. Turn left (west) on 2nd Street, which becomes Saratoga Road at the corner with De Bruyn Avenue. Saratoga Woods Preserve is on the left 2.5 miles (4 km) from the corner with De Bruyn. There may be maps at the trailhead kiosk. Take Wood Nymph Way from the north end of the parking lot. Go right at the junction with Bent Tree Trail. About ten minutes from the car you will cross an overgrown airstrip, now a dirt road. Continue on the trail across the airstrip, dropping into a minor depression. The Waterman erratic is 200 feet (60 m) beyond the airstrip, hidden in the deep woods.

The gigantic Waterman erratic is so large that it is difficult to take a photograph that does it justice.

Lake Stevens Erratic, Snohomish County

The monstrous Lake Stevens erratic is the largest known in the state. It is 34 feet (10 m) tall and 78 feet (24 m) long. Its circumference is 210 feet (64 m). In fact, it might be the largest erratic in the United States. It only became known to geologists in the summer of 2011, when two Lake Stevens police officers with a penchant for geology reported it.

The erratic is partially buried in glacial till and outwash, which banked around it after the ice was done carrying the large rock around. Postglacial stream erosion has removed some of the sediment, so the erratic is perched on a slope above a gully. Not much of it is visible from the street, although it does look pretty long. Only when you walk the narrow path to the vertical east face does it become apparent that you are in the presence of a really huge rock.

Unfortunately, because of the trees, brush, and slope of the ravine behind the high side of the erratic, it is difficult to get a good perspective of the size of this rock, or a photograph that does it justice.

Like the Waterman and Coupeville erratics, this one is also greenschist. The rock of this giant erratic was likely erupted onto the seafloor of the Pacific Ocean and then carried eastward and subducted beneath North America. Don't puzzle too long over the gray

The Lake Stevens erratic.

GETTING THERE: From I-5 in Everett head east on US 2 for 2 miles (3.2 km). Take the 20th Street SE exit straight ahead and go about 1 mile (1.6 km) to the stoplight at 83rd Avenue. Turn left (north). After 1.3 miles (2.1 km) you'll pass 1st Street SE. The erratic is in a small park maintained by the people of the neighborhood, on the right before the cul-de-sac ending 83rd Avenue. There is an informal trail around the south side to the erratic's dark, looming east face.

cement-like material smeared on the backside of the erratic; it really is cement, applied to cover footholds and handholds to discourage climbers. You'd think that the slimy wet rock would be sufficient to do that.

The green color in greenschist is from the low-grade metamorphism of iron- and magnesium-rich minerals, which turned them into chlorite.

It is not possible to say with confidence where the erratic began its journey on or within the great Puget Lobe. There is similar-looking greenstone of the Fidalgo Complex exposed at the north tip of Whidbey Island, 37 miles (60 km) northwest of Lake Stevens; this is the probable source for the Coupeville and Waterman erratics. However, the Lake Stevens erratic seems a little too far to the east to have been carried from the Fidalgo Group exposure to its current location by a southward-flowing glacier. Rocks in the North Cascades, especially around Mount Shuksan, as well as up the Fraser River, are a more likely source area.

Prior to discovery of the Lake Stevens monster, the largest known erratic in the United States was the Madison Boulder in New Hampshire. This granitic monolith measures 83 feet (25 m) in length, 23 feet (7 m) in height above the ground, and 37 feet (11 m) in width. One thing we don't know about any of these erratics is how much of each is still buried in the ground. It appears that a large part of the Lake Stevens monster remains buried.

Wedgwood Erratic, Seattle

This erratic, yet another Big Rock, is 19 feet (6 m) tall and 75 feet (23 m) in circumference. Like the monstrous Lake Stevens erratic, it is greenschist and could have come from the same source, wherever that may be. The rock, in the Wedgwood neighborhood, has played a role in Seattle's history. The oldest written descriptions of the rock date to 1881, with an account of a Fourth of July picnic at and on the easily climbed boulder. At the time, the erratic was called Lone Rock, and it was surrounded by forest on William Weedin's farm.

Winlock Miller owned the land into the 1940s and kept it undeveloped. Albert Balch bought the property from Miller and began to develop the residential district. Balch promised Miller he would make a park around the rock. When he failed to do so, a neighborhood

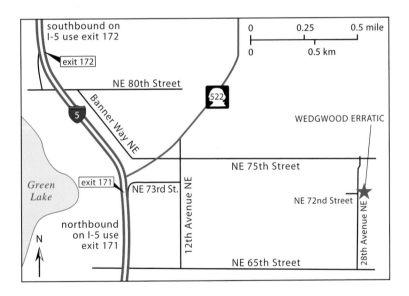

GETTING THERE: This urban erratic, also known as Big Rock, is in a neighborhood park between Sand Point and Green Lake. Southbound on I-5, take exit 172 to NE 80th Street. Turn left (east) and cross over the freeway, then promptly turn right on Banner Way NE, which becomes NE 75th Street. If northbound on I-5, take exit 171. Just before the ramp merges with Washington 522, veer right onto NE 73rd Street. At the intersection with 12th Avenue NE, turn left and travel one block to NE 75th Street. Once you reach NE 75th Street, head east for about 1 mile (1.6 km) and turn right onto 28th Avenue NE. The erratic is on the east side of the street at the intersection with NE 72nd Street.

The Wedgwood erratic.

group sought Seattle City Council support to purchase the land as a city park in 1946. That also failed; however, today the land immediately around the erratic is maintained by the neighborhood as de facto open space.

Seattle's famed Mountaineers Club had an arrangement to use the rock for climbing practice in the early 1900s. According to a neighborhood newsletter, Lloyd Anderson, one of the founders of Recreational Equipment Incorporated (REI), taught rock climbing here. Among the young climbers who learned basic climbing methods at Big Rock were Fred Beckey and Jim Whittaker. The former became the foremost first-ascent pioneer in the Cascade Range; Whittaker was the first American to climb Mount Everest.

The rock climbing ended, legally anyway, in 1970. In the sixties, the top of the rock was a popular hangout for hippies. Locals were concerned that the longhairs were using drugs, harassing residents, and leaving trash. Residents even thought the rock's vista was being used to case the neighborhood for burglaries. The Seattle City Council passed an ordinance in October 1970 prohibiting climbing on Big Rock. The fine is $100.

Lake Lawrence Erratic, Thurston County

The landscape near the Lake Lawrence erratic is a mix of open fields, forests, and rural housing developments. It seems pretty featureless. That's why the squarish, garage-sized granodiorite erratic suddenly appears as an anomalous surprise in a clump of trees right beside 153rd Avenue SE. The side of the 17-foot-high (5 m) rock facing the road is shaded and largely covered in moss. If you walk around to the back side and squeeze up against the fence, you can get a good look at a moss-free—and maybe even sunny—exposure.

Granodiorite is a plutonic rock, an igneous rock that cooled within the crust. Plutonic rocks consist entirely of mineral crystals, which in this rock are generally 0.1 inch (3 mm) across. Granodiorite is characterized by its distinctive salt-and-pepper color. If you can't find a small chip lying on the ground, you'll have to stick your nose right up to the rock to examine the crystals. Use your hand lens. The

The moss-draped Lake Lawrence erratic sits right on the edge of the road.

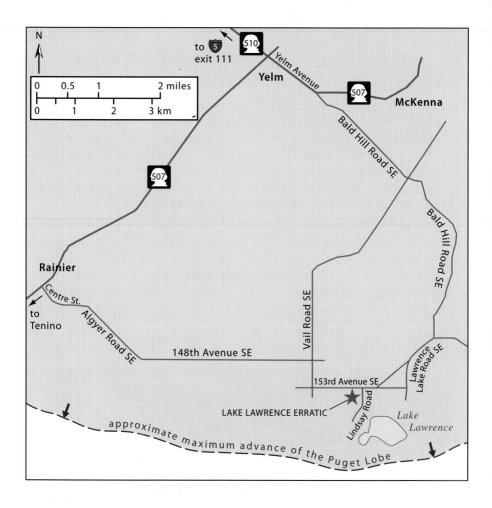

GETTING THERE: From Olympia and points north, take exit 111 off I-5. Head east on Washington 510, which becomes Washington 507 in the center of Yelm, about 13 miles (21 km) from exit 111. Take Washington 507 southeast out of Yelm. One mile (1.6 km) from the town center turn right on Bald Hill Road SE. After 5 miles (8 km) turn right on Lawrence Lake Road SE. After another 1 mile (1.6 km) turn right on 153rd Avenue SE. The erratic is on the left in 0.9 mile (1.4 km), just beyond the fire station at Lindsay Road SE.

From points south, leave I-5 at exit 88 (Rochester) and follow Old Highway 99 SW for 7.8 miles (12.6 km) to Tenino. Continue on Washington 507 for 9 miles (14.5 km) to Rainier. Turn right (east) on Centre Street, which becomes Algyer Road SE and then 148th Avenue SE. Five miles (8 km) from Rainier turn right on Vail Road SE, and after 0.5 mile (0.8 km) turn left on 153rd Avenue SE. The erratic appears suddenly on the right in a little under 0.7 mile (1.1 km).

Parking along 153rd is problematic, as there are no formal parking spots. Do not park in front of the fire station; you might be able to park across from the fire station in a pull-out. You will need to walk along the shoulder about 200 feet (60 m) to reach the erratic.

The dark biotite and light-colored quartz and plagioclase feldspar in the Lake Lawrence erratic are characteristic minerals of granodiorite.

dark crystals take the form of stacks of thin sheets; they are flaky looking. This is biotite, an iron-rich member of the mica family of minerals and a very common constituent of plutonic rocks. It has a bronze-brown color, though it may appear silvery if you look at the flat, shiny surface of one of the sheets. The light-colored minerals are whitish plagioclase feldspar and grayish quartz.

Granodiorite erratics are probably the most typical (and easily recognized) erratics in the Puget Lowland because this rock is common in the British Columbia Coast Mountains far to the north, and because it is durable and withstands the bump and grind of glacial transport. These erratics may be polished smooth and may have scratches and even deep glacial striations ground into the rock by the rigors of glacial transport. (I didn't see any striations on this erratic, but it is too mossy to see very much.) This big boulder is pretty symmetrical, and each side is about 17 feet (5 m) long. The volume is somewhere around 4,900 cubic feet (140 m³), and it weighs about 820,500 pounds (372,200 kg), or 410 tons (370 metric tons). For comparison, the maximum loaded takeoff weight of a Boeing 747 jetliner is 975,000 pounds (442,250 kg).

The erratic is significant because it lies on the lumpy moraine topography marking the maximum southern advance of the Puget

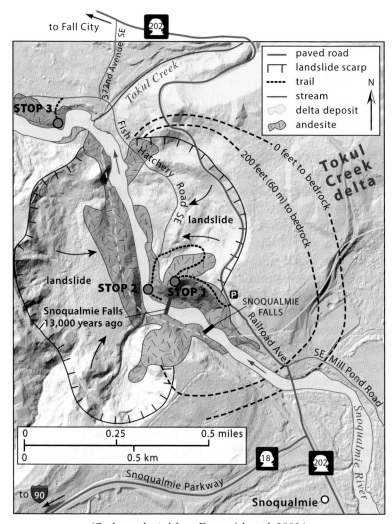

(Geology adapted from Dragovich et al. 2009.)

GETTING THERE: Stop 1 uses the viewpoints high above the Snoqualmie River canyon at Puget Sound Energy's Snoqualmie Falls Park. From the east, south, or west, take exit 25 off I-90. Head north on Washington 18 (Snoqualmie Parkway) for 3.7 miles (6 km). Turn left at Washington 202 (Railroad Avenue), cross the Snoqualmie River on a steel bridge just above the falls, and arrive at the park in 0.8 mile (1.3 km). From the north, leave US 2 in Monroe. Turn south on Washington 203. After traveling along the rural Snoqualmie River valley for about 24 miles (39 km), merge with Washington 202 (SE Fall City–Snoqualmie Road) at a roundabout in Fall City. Follow signs toward the falls, which are another 3.9 miles (6.3 km). Park in the big lot on the east side of the road, across from the falls; the lot on the west side of the road is usually jammed. Snoqualmie Falls Park is free. Stops 1 and 2 are open from dawn until dusk.

the lake, crossing a narrow part of the ancestral Snoqualmie River valley to bank against the thicker and higher pile of older sediment comprising the hills immediately west of the falls.

The Tokul delta buried older volcanic rocks, now exposed at Snoqualmie Falls and along the river for a short distance north. The volcanic rocks at the falls are 22.7 to 18 million years old; they include a cap of andesite lava and a thick underlying layer of tuff breccia, a volcanic rock consisting of poorly sorted fine-grained ash and large

Snoqualmie Falls flowing at 1,800 cubic feet (51 m³) per second.

Stop 2 is on the riverside at the Lower Falls Viewpoint. A 0.4-mile-long (0.6 km) trail descends 280 feet (85 m) to the river from the north end of the viewpoints at stop 1. Stop 2 can also be reached by road. Go north 1.5 miles (2.4 km) on Washington 202. Turn left at 372nd Avenue SE, and then left again onto SE Fish Hatchery Road, which leads to the lower parking area. Take the trail and boardwalk southward 0.2 mile (0.3 km) to the end.

Stop 3 is on the river downstream from stop 2. If you hiked to stop 2, return to the upper parking area and follow the road directions to stop 2. At the intersection of 372nd Avenue SE and Fish Hatchery Road, turn right, downriver. If you drove to stop 2, return to the intersection and go straight through on Fish Hatchery Road SE. Go downstream about 520 feet (160 m), passing the Department of Fish and Wildlife river access at the road junction, where there are outhouses, and park at a small pullout on the river side of the road.

angular blocks. The relatively soft rocks of this deposit may have been blasted out of a volcanic vent during an eruption, or they may have surged down a volcano's flanks as pyroclastic flows of glowing hot ash and rock fragments. The blocks composing the breccia are up to 9 feet (2.7 m) across, too large to travel very far through the air, indicating that a volcano grew very close to the modern falls in the Miocene epoch. These volcanic rocks are only exposed in the river valley near the falls, where they have been exhumed by the Snoqualmie River. They may be more extensive, but a thick blanket of glacial and interglacial sediment covers them.

As the Puget Lobe receded at the end of the Pleistocene, and then Glacial Lake Bretz and its arm filling the Snoqualmie River valley drained to the Strait of Juan de Fuca, the river reestablished itself in the valley. The nascent river encountered the blockage of the Tokul delta and flowed around the toe of the delta at its lowest point, where the delta butted against the hills west of the falls.

The Snoqualmie rapidly cut down into the thick, gravelly deposits of the delta, trenching a V-shaped gorge. There was almost certainly a set of rapids at that time as the river tried to establish a gentler gradient in its course to the sea. When it had cut deeply enough into the delta, though, the river encountered the buried andesite lava and pyroclastic deposits; it began to cut into these rocks, too. By around 13,000 years ago the river was confined within a bedrock gorge it had cut. At the head of the gorge were the falls.

Stop 1 is at the viewpoint next to Salish Lodge, on the brink of the canyon. In the spring or during floods, the river's discharge is high, and the falls are particularly impressive. Even on a sunny day you may get wet from windblown spray.

The river plunges in free fall over the cliff from a notch, or knickpoint, eroded 100 feet (30 m) down into the andesite. Salish Lodge sits on the brink of the cliff 490 feet (149 m) above sea level. The current elevation of the rock at the knickpoint is 390 feet (119 m); the water falls 268 feet (82 m) to the roiling plunge pool beneath. The size of the falls depends on the river's volume, which is greatest during a Pineapple Express, when warm rain falls on the snow-packed Cascades. Other good times to see high flows are when the river is swollen with snowmelt in early summer. The record discharge since 1930 (the first year of stream gauge records) occurred on January 9, 2009, when 60,492 cubic feet (1,713 m³) roared over the cliff every second. How much water is this? An Olympic swimming pool holds about 88,000 cubic feet (2,500 m³). More than half of an Olympic pool's worth of water passed over the falls every second. The average discharge is 501 cubic feet (14 m³) per second.

Peregrine Viewpoint is perched on the canyon edge at the north end of the paved walkway; it is the second viewpoint north of Salish Lodge. This was the location of the falls about 13,000 years ago, when they first formed. Here you would have been standing on the brink as the water thundered over the cliff. There was a continuous wall of rock across the river from Peregrine Viewpoint to the high cliffs on the other side. From this point, Snoqualmie Falls has cut a channel backward about 1,050 feet (300 m) through the stack of volcanic rocks at an estimated recession rate of 1 inch (2.5 cm) per year.

Blocks of porphyritic andesite from the lava are arranged at various places around the upper overlooks. White plagioclase up to 0.2 inch (5 mm) across is the dominant mineral, as it usually is in Cascades andesite. You may spot very dark green or black augite and perhaps honey-brown hypersthene. Flaky bronze-black biotite is present but uncommon.

The geology in this stretch of the Snoqualmie River dooms the falls to a short life. The river's elevation at the bottom of the falls is 140 feet (43 m). The top of the falls is 390 feet (119 m). Recent work has shown that the volcanic rocks disappear below a thick accumulation of alluvial sediment only 600 feet (180 m) upstream of the fall's brink; the highway bridge 450 yards (400 m) above the falls is built on river sediment, not bedrock. About 300 yards (270 m) above the falls the buried bedrock is 200 feet (60 m) below the surface, only 50 feet (15 m) higher than the river at the bottom of the falls. The volcanic rocks continue to get deeper toward the southeast. So, the river will migrate backward through the lava for another 600 feet (180 m) until it hits the alluvium; at its current recession rate, that will take 7,200 years, a geologic eyeblink. It will probably be even faster since most of the volcanic rock above the falls is breccia, not the more massive and resistant lava. The falls will be replaced with a short series of rapids for as long as the river flows over volcanic rock, but the river will be at a much lower elevation than it is today upstream of the falls, and it will be flowing through a V-shaped gorge lined with soft Pleistocene sediment.

Head down to stop 2. If you take the road, you will descend a steep road grade across the face of the Tokul delta. This stretch is very prone to landslides. The road crosses Tokul Creek, where you can get a quick glimpse of the delta deposits the creek has cut through.

You can better appreciate the knickpoint from the viewpoint at the end of the paved trail and boardwalk. Be prepared for spray. The lower part of the rock cliff is tuff breccia that was erupted from a nearby volcanic vent. Once the river cuts downward through

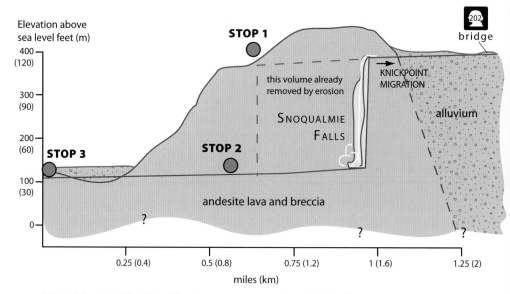

The knickpoint of the falls will migrate upstream and reach thick alluvium relatively quickly, causing the demise of Snoqualmie Falls. The area outlined in red is the amount of rock that has been removed in the past 13,000 years. Vertical scale is greatly exaggerated.

Big blocks of volcanic rock are surrounded by a fine-grained matrix in the tuff breccia below the falls. These are on the west bank. The Snoqualmie River easily erodes this rock. —Courtesy of Joe Dragovich, Washington State Department of Natural Resources

the andesite capping the cliff, it will more easily remove the tuff beneath. The falls are free falling from the overhanging lip because of this differential erosion. You can see the breccia on the opposite side of the river (private property), especially if you bring binoculars. Look for large blocks in the finer matrix.

Snoqualmie Falls from stop 2 in late spring. Discharge was 5,500 cubic feet (155 m³) per second. —Courtesy of Mark Kiver

Stop 3 allows close examination of the andesite lava. From the parking spot, take the right-hand trail and walk 25 yards (20 m) toward the river. Descend a slope of sand and cobble alluvium resting on top of a 10-foot-high (3 m) wall of andesite lava. Beneath the lava a ledge of breccia extends well out into the river (unless water levels are high). A basal breccia layer like this is typical of andesite lava flows. As the lava advances, the front of the flow breaks up into blocks, which are then overridden by the lava flow. The best breccia exposures are close to the water. This breccia should not be confused with the tuff breccia exposed beneath the falls; though composed of similar rock, they were deposited in a different manner. This is the farthest north that Miocene volcanic rocks are exposed along the river. From here the river flows placidly (except during floods) to its junction with the Skykomish River and then to a rendezvous with Puget Sound at Everett.

At stop 3, breccia at the base of the andesite lava flow forms a ledge projecting into the river.

14

A Tour of Downtown
Seattle Building Stone

Structural and ornamental building stones (also called dimensional stone) used on and in buildings in downtown Seattle are also found on world-famous buildings, including the Empire State Building and the Roman Coliseum. The natural geology of downtown is buried by cement and asphalt "urbanite" — man-made rock. Below that thin skin, up to 35 feet (11 m) of redistributed natural sediment was unceremoniously dumped as fill. Much of the pre-urban topography downtown has been leveled or at least regraded, but the city's east-west streets still rise steeply uphill from the waterfront.

Building stone from around the world as well as Seattle's backyard is found in the canyons of downtown. The stones are a great introduction to petrology, the identification of rocks and their origins. They are found on building facades and in fancy lobbies. Some are at first (and last) glance rather prosaic, but others are beautiful

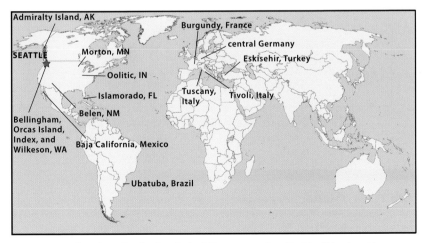

Source locations for the Seattle building stones described in this vignette.

209

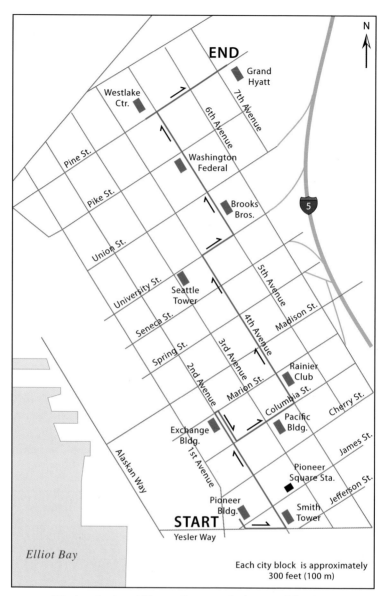

N

Grand
Hyatt

END

Westlake
Ctr.

6th Avenue

7th Avenue

Pine St.

Washington
Federal

Pike St.

Brooks
Bros.

5

Union St.

University St.

Seattle
Tower

5th Avenue

Seneca St.

Madison St.

Spring St.

4th Avenue

2nd Avenue

3rd Avenue

Marion St.

Rainier
Club

Columbia St.

Cherry St.

Exchange
Bldg.

1st Avenue

Pacific
Bldg.

James St.

Alaskan Way

Pioneer
Square Sta.

Jefferson St.

Pioneer
Bldg.

Smith
Tower

START

Yesler Way

Elliot Bay

Each city block is approximately
300 feet (100 m)

The tour begins at Pioneer Square at the bottom of the map.

GETTING THERE: This walking trip begins at Pioneer Square at the intersection of 1st Avenue, Yesler Way, and James Street in the heart of Seattle's historic district. The one-way distance is 1.4 miles (2.3 km) on city sidewalks. Elevation gain is about 100 feet (30 m). The leg from 2nd Avenue to 4th Avenue on Columbia Street is the steepest, gaining around 66 feet (20 m) in two blocks.

Street parking is reminiscent of combat, and garage parking is expensive. Metro Transit provides convenient service to downtown. The most convenient transit access is the

and exotic. There are rocks with fossils, rocks from deep in the crust, rocks from distant and long-gone seafloors. Urbanite predominates. Examples include brick and concrete. Some of this faux rock cleverly masquerades as natural rock and requires careful investigation to penetrate the disguise.

Stone? Rock? What's the difference? Rock is a natural aggregate of minerals (igneous rocks and most metamorphics), other rock fragments (sedimentary rocks and some metamorphics), or organic material (for example, coal). Stone is rock used in construction or art, or it can be one of the larger fragments in a sedimentary rock, such as pebbles in conglomerate. The more practical definition used by stone carvers is, "If there's a price tag on it, it's stone."

Rocks used for buildings and ornamental purposes often carry a common trade name that bears little or no relation to their geology. While granite has a specific petrologic definition, the term is used very loosely with regard to dimensional stone. Commercial websites often offer polished countertop "granite" slabs that are actually marble or gneiss. On the same website, you may find granite sold as "marble." Other building stones have widely used informal names based on color or source. Consider Green Ubatuba granite, which may now be quarried from around the world; geologically the rock sold under this trade name is metamorphic, so it's not the actual granite quarried at Ubatuba, Brazil. Chuckanut sandstone is another example. It is sandstone, and it is from the Chuckanut Formation, but the Chuckanut includes shale and conglomerate too.

You won't find Rainbow granite or Chuckanut sandstone mapped on a geologic map, though you will find Morton Gneiss and Chuckanut Formation. Confused? So am I, sometimes. The table accompanying this vignette should help you sort out the official formation names and appropriately used rock type names from misappropriated usage in trade names.

Pioneer Square Station on James Street between 2nd and 3rd Avenues. Many bus lines run along 3rd Avenue; hop off at Cherry or James Street, only a couple of blocks from Pioneer Square. Seattle avenues run roughly north-south. Seattle's east-west streets may run steeply uphill or downhill, but avenues are level. This trip is mostly along the avenues. The tour goes in and out of building lobbies. The open times are as up-to-date as possible at the time of writing. You may need to ask permission if you wish to take photos. Interior photography is not permitted in the Westlake Center.

LOCATION	ROCK TYPE	GEOLOGIC UNIT	TRADE OR POPULAR NAME	STRATIGRAPHIC AGE	RADIOMETRIC AGE (years ago)	QUARRY LOCATION
Pioneer Building	sandstone	Chuckanut Formation	Chuckanut sandstone	Eocene	50 million	Bellingham and Orcas Island, WA
Smith Tower	granodiorite	Index batholith	Index granite	Eocene	35 million	Index, WA
	marble	none	Tokeen marble	Silurian	430 million	Admiralty Island and vicinity, AK
	limestone	Heceta Formation	Aphrodite marble	Silurian	430 million	Admiralty Island and vicinity, AK
	onyx marble	none	Pedrara marble	Pleistocene?	not kown	Baja California, Mexico
Exchange Building	migmatite gneiss	Morton Gneiss	Rainbow granite	Archean	3.52 billion	Morton, MN
	limestone	La Spezia Formation	Portoro	Late Triassic	237–201 million	Tuscany, Italy
Pacific Building	travertine	Lapis Tiburtinus Formation	Tivoli travertine	Pleistocene	115,000–30,000	Tivoli, Italy
Rainier Club	limestone	Salem Formation	Indiana limestone	late Paleozoic	359–300 million	Oolitic, IN
Brooks Brothers	sandstone	Carbonado Fm.	Wilkeson sandstone	Eocene	45 million	Wilkeson, WA
Westlake Center	limestone	Comblanchien Fm.	Comblanchien ls.	Jurassic	160 million	Burgundy, France
	limestone	Key Largo Fm.	Florida limestone	Pleistocene	130,000	Islamorada, FL
Westlake Station	migmatite gneiss	Morton Gneiss	Rainbow granite	Archean	3.52 billion	Morton, MN
	travertine	Mesa Aparejo travertine	New Mexico travertine	Pleistocene	≥321,000	Belen, NM
	charnockite	none	Ubatuba and variants	Cambrian	500 million	Ubatuba, Brazil
Grand Hyatt Hotel	limestone	Treuchtlingen Fm.	Treuchtlingen marble	Jurassic	155 million	central Germany
	onyx marble	none	honey onyx	Pleistocene?	not known	Eskisehir, Turkey

A summary of dimensional stones visited on the tour.

The columns on the Pioneer Building are blocks of Chuckanut Formation sandstone.

The Pioneer Building (600 1st Avenue) is a beautiful National Historic Landmark. It was completed in 1892 and was among the first "fireproofs" to rise following the devastating Great Seattle Fire of 1889. Most of the exterior cladding is red brick, but the stack of disklike stones in the tan columns rising the full height of the building's front are distinctively rough-cut sandstone from the Eocene-age Chuckanut Formation (see also vignettes 20 and 21). Quarries were on Orcas Island and the shore of Chuckanut Bay near Belling-ham, both convenient for supplying building stone quickly via ship to reconstruct the city. Chuckanut sandstone blocks are often left in a rough-finished state, with scalloped surfaces. You are unlikely to mistake it for other sandstones due to the finish and the color; you'll spot it throughout the field trip on other buildings, especially as foundation blocks. Chuckanut sandstone was also shipped to San Francisco to rebuild that city after the 1906 earthquake and fire.

Some of the sandstone blocks on the ground floor windowsills at the corner of James and Yesler have begun to exfoliate, or flake, along subtle bedding planes. This type of weathering develops when water enters the stone and dissolves the natural silica cement binding sand grains together. Water really accelerates exfoliation because it expands when it freezes. Look closely to see the close-set layering in the rock, parallel with the windowsill.

Stroll east on Yesler to 2nd Avenue and crane your neck to see the white Smith Tower across the street. This is among the most recognizable of Seattle's icons. When completed in 1909, the 462-foot (149 m), thirty-eight-story building was the tallest building west of Chicago, remaining so until 1931, and it was Seattle's tallest building until 1969. You can get a good view of the whole building from the parking lot across 2nd Avenue.

The exterior above the second floor is blinding white terra-cotta. *Terra-cotta*, Italian for "baked earth," is an ancient form of urbanite. Clay is baked in a kiln and glazed for waterproofing. Terra-cotta is less dense and lighter than stone and is commonly used on the upper floors of buildings in Seattle as cladding, the thin exterior finishing material that is applied over brick, wood, or metal walls.

The stone on the first two floors are salt-and-pepper-speckled Eocene-age granitic rock quarried from the Index batholith at the town of Index, less than 40 miles (64 km) east of Everett on US 2. The Index batholith has a range of compositions, from quartz diorite to granite. The rock at Index, popularly called Index granite, is granodiorite, as there is not enough quartz to classify it as granite. Index granodiorite is found on the facades of many Seattle buildings, and on buildings and sculptures throughout western Washington. Dark minerals in the Index are hornblende and biotite mica; the white minerals are largely plagioclase feldspar with some quartz. The abundant rounded blobs of darker rock up to 3 inches (8 cm) across are intrusive enclaves, mafic magma of a different chemistry that intruded the batholith before it solidified. Mafic rocks have higher melting and solidifying temperatures than granitic rocks. As long as the enclaves were molten, they heated the Index magma and slowed its solidification.

Go in the 2nd Avenue entrance. The building is open seven days a week. In the elevator lobby you'll see two different stones on the polished walls. The wavy black-and-white stone is Silurian-age Tokeen marble, quarried on Marble Island in southeast Alaska. The marble formed when intruding magma baked and metamorphosed 430-million-year-old limestone, obliterating any trace of fossils in the original limestone.

Tokeen marble was among Alaska's main stone exports in the early twentieth century. Getting the marble from the quarry on the isolated island to the Smith Tower was quite a feat. Rough-cut

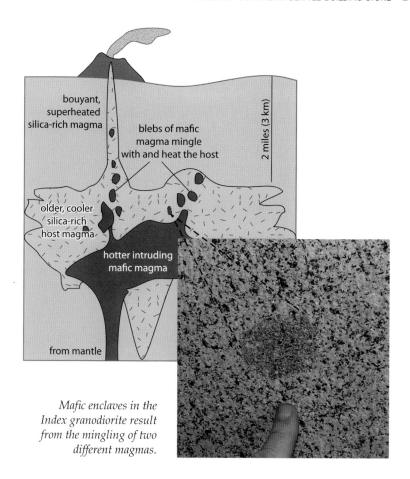

bouyant,
superheated
silica-rich magma

blebs of mafic
magma mingle
with and heat the host

2 miles (3 km)

older, cooler
silica-rich
host magma

hotter intruding
mafic magma

from mantle

*Mafic enclaves in the
Index granodiorite result
from the mingling of two
different magmas.*

blocks weighing 10 tons (9 metric tons) or more were cut by hand- and steam-operated saws, loaded on gravity-powered flatcars, and run down to the beach. They then were hoisted aboard barges and shipped to a finishing mill in Tacoma to be sawn and smoothed. The marble appears in buildings all over America, including the legislative building in Athens, Georgia, and the rotunda of the Washington State Capitol. Tokeen marble is found in other buildings in Seattle, including the King County Courthouse (516 3rd Avenue) and the Hoge Building (705 2nd Avenue).

The beige stone panels that look like a geologist's Rorschach test are onyx marble from quarries of the Pedrara Onyx Company at El Marmol (Spanish for "marble"), in the northern part of Baja California. Don't confuse this calcium-rich sedimentary rock with the silica-rich, sedimentary onyx, nor with the metamorphic marble. Onyx is a black-and-white-banded agate, a variety of quartz without crystalline structure. Marble is a calcium-rich metamorphic rock. Like true marble, onyx marble is principally calcium, but unlike marble it is not metamorphic. Onyx marble forms as calcium carbonate precipitates out of water in pools and hot or cold springs. The bands of colors that differentiate onyx marble from two other spring-deposited calcium carbonate rocks, travertine and tufa, derive from oxidized impurities in the calcium deposit, especially iron and manganese. Calcium is a much softer mineral than quartz, so onyx marble is more easily cut than many building stones. It also takes a glossy, smooth polish. These features make onyx marble popular for use as thin sheets covering walls. The vertical brown bands of the panels were originally deposited horizontally in springs. The panels were cut along these bands and rotated 90° from the original orientation of the layers. Onyx marble panels cover the walls in the elevator lobby. The more or less structureless grayish panels are also onyx marble, but they were cut parallel to the color bands so you don't see them.

Since most onyx marble comes from El Marmol, it is also known as Mexican onyx. The quarries in Baja operated from 1893 until 1970. Demand declined with the advent of cheap, lightweight, abundant plastic. The rock is probably young, perhaps Pleistocene in age.

Go back to the entryway and look at the stone along the stairs going down. Here are more of the grayish onyx marble panels. At the foot of the first flight of steps the marble is abruptly replaced. Myriad nightmarish shapes float across a beige background. The dark ones leap out at you, but there are also ghostly white forms. These fossils are preserved in limestone from the Silurian period, 444 to 419 million years ago. This is when plants and animals first emerged into terrestrial environments, but marine life, such as the examples on this wall, represented the majority of Earth's species. Most of these fossils are pelecypods, mollusks with two valves, or shells, similar to today's clams and oysters. There are also large horn corals, which in these slabs are asymmetric, elongate, and generally white. These once-abundant animals died out 251 million years ago at the close of the Permian period, during Earth's greatest mass extinction event. Horn corals were solitary animals living in curved, hornlike, calcareous exoskeletons with rough walls. They lived on massive coral reefs, which were abundant in the Silurian.

The source of these polished slabs is not certain. Silurian sedimentary rocks are rare in North America. However, limestone from the Silurian Heceta Formation, with a similar fossil assemblage, was quarried from islands off the shore of Prince of Wales Island, about 3 miles (5 km) from the Tokeen marble quarry. Given the presence of Tokeen in the Smith Tower, and the proximity of the Heceta and

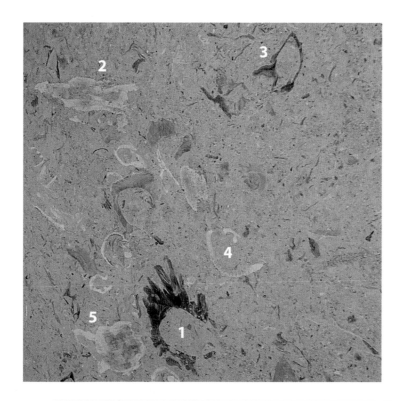

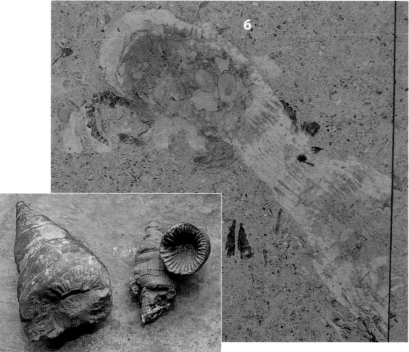

Fossils in the Heceta Formation limestone: 1 is possibly a microbial mat, 4 inches (10 cm) wide; 2 through 4 are bivalve fossils, likely clams but possibly brachiopods; 5 could be an oblique section through a snail; 6 is a nicely preserved, 8-inch-long (20 cm) horn coral. Inset: horn corals of Paleozoic age (Western Washington University collection).

Tokeen quarries in Southeast Alaska, it seems reasonable that the source of these fossil-rich slabs is the Heceta Formation. Much of the Heceta was deposited in shallow marine reefs. Sculptors call limestone from the Heceta Formation Aphrodite marble, another classic example of a misnomer.

From the Smith Tower, walk north on the west side of 2nd Avenue to the twenty-two-story art deco Exchange Building (821 2nd Avenue). The bronze plaque on the wall near the southwest corner of 2nd and Marion is mounted on two building stones that couldn't be more different. The pale stuff beneath the upper part of the plaque isn't a geologic product, but a man-made, cement-based urbanite. Cement is mixed with lightweight, angular pebbles of pale pumice and other stones and cast as slabs for building walls. Porous, highly vesicular pumice reduces the weight. No architect or builder wants to call such prosaic slabs concrete, so a catchier name is used in the industry: Romanite Stone, a tribute to its origin.

The Romans invented cement and concrete. They started by crushing and heating rock rich in calcium carbonate, such as limestone or marble, in a kiln. The carbon and two-thirds of the oxygen were driven off as gases, leaving calcium oxide, or quicklime. The Romans mixed the quicklime with water to make a strong, chemically bonded paste—hydrated quicklime—and cemented pebbles or crushed rock in an artificial stone we call concrete, which is simply cement (quicklime) plus aggregate.

The pink-and-black-swirled rock beneath the plaque is one of the most celebrated building stones in America, the Morton Gneiss. Commercially known as Rainbow granite, this is the oldest building stone you'll find in the country, and perhaps the entire world. This rock, from the Archean eon, is an amalgamation of different rocks mixed together at different times and by different events, but the oldest components are 3.524 billion years old, more than three-quarters the age of our planet. The rock is quarried at Morton in southwestern Minnesota. The ancient rocks in that region are the remnant of a long-vanished, early version of the North American Plate.

The Morton Gneiss is migmatite, a mix of metamorphic gneiss and plutonic granite. Gneiss is a coarsely crystalline, foliated metamorphic rock resulting from extreme heat and pressure acting on other rocks, which deform but do not quite melt. The oldest portion of the migmatite started out as magma with tonalite composition, similar to the Smith Tower's Index granodiorite. Millimeter-size zircon crystals in the tonalite provide the staggeringly old ages for this part of the Morton Gneiss. At an unknown time, the tonalite mixed with basalt, the prominent black portion of the rock. This basalt could be from an older volcanic crust intruded by the tonalite, or it could be a younger magma that mingled with and solidified within the cooler tonalite. Alas, the basalt contains no zircons, so we can't measure its age.

Much later, 2.68 billion years ago, granitic magma rich in pink-and-gray potassium feldspar intruded the mix. A mighty mountain range must have existed at this time because the intrusion occurred at such great depths in the crust that the basalt, despite its high melting point, was weakened by immense pressure and stretched like taffy. Minerals were aligned by flow and pressure to give the rock its coarse-grained and foliated texture. The granitic magma flowed between the basalt blocks, rotating and swirling them like rafts on a gooey pink river. Thin stringers of granite and softened basalt were folded together like chocolate-chunk cake batter. The rock cooled into the intricate mix of pink granite and black basalt we see today. A few billion years of uplift and erosion brought the gneiss to the surface for the burly quarrymen to cut into blocks.

Walk a few steps down Marion and enter the Exchange Building (closed weekends). The dark, polished zebra-striped limestone slabs covering the walls seem to swallow the light in the dim lobby.

Highly polished Morton Gneiss adorns the 2nd Avenue side of the Exchange Building. The dark part of this migmatite rock is basalt, heated and deformed by later intrusion of pink granitic magma.

This psychedelic rock is limestone from the La Spezia Formation, from the Tuscan coast of northwestern Italy. It is known in the stone industry as Portoro limestone or Portoro marble. Similar rock is now exported from China and marketed as Italy Portoro.

Portoro limestone was deposited in shallow water on the margin of the Tethys Ocean as calcium-rich mud 237 to 201 million years ago, during the Late Triassic. The Tethys Ocean expanded as the supercontinent Pangaea rifted and split into two gigantic tectonic plates, Gondwana and Laurasia. The Mediterranean Sea survives as a remnant of this once extensive ocean. The tiny grains that rained onto the floor of the Tethys were mostly the partially dissolved skeletons of microscopic red algae. The limestone's black color is typical of sediment containing poorly decomposed organic material, deposited when low oxygen conditions existed, and the creamy yellow layers represent intervals when more oxygen was available and the organics decomposed more thoroughly. From the appearance of the limestone, oxygen levels in this part of the Tethys Ocean fluctuated rhythmically.

The limestone was deformed 180 million years ago, during the slow-motion collision of Africa and southern Europe, which is closing the Mediterranean Ocean today and will lead to the final demise of the Tethys. Collision caused compression, which in turn caused slippage between layers in the limestone. This bedding-parallel shearing occurred within the pale strata, which were weaker than the black rock. The shearing tore off bits of the brittle black layers and left pieces embedded within the pale-yellow bands. The rock also broke across bedding planes; these fractures filled with calcite to produce the white crosscutting veins, best seen near the lobby's elevators.

The stonecutters at the Italian quarry sliced the rock across the bedding to produce the Portoro's distinctive zebra banding. The width of the color bands and the shapes of dark clasts in stacked pairs of slabs are nearly identical but reversed. They came from adjacent slabs sliced from the original quarry blocks. The Exchange Building's masons cleverly matched the reversed slabs of polished limestone to create the striking chevron pattern.

Return south one block on 2nd to Columbia and walk up the hill. Cross 3rd to the Pacific Building (720 3rd Avenue). The stone with the wavy vertical patterns facing Columbia is travertine of the Lapis Tiburtinus Formation, quarried at the famous Italian quarries in Tivoli, east of Rome. Travertine forms in springs. When hot water passes upward through calcium-rich rocks, it dissolves the calcium, precipitating it as a very fine-grained, homogeneous mud. The hot

Yellowish beds in the Portoro limestone contain blocks torn loose from the dark limestone beds by bedding-parallel shearing.

springs at Yellowstone National Park are famous examples of places where travertine is being deposited today. The Tivoli travertine, fairly soft and easy to quarry, was used to build the Coliseum in Rome as well as the Trevi Fountain and the columns of Saint Peter's Basilica. The Tivoli travertine is only 115,000 to 30,000 years old and is almost surely the youngest building stone you'll find downtown.

The layering was originally horizontal, but the stone slabs are standing on end. Elongated cavities characterize travertine. These mark where algae lived on the muddy calcium carbonate floor of a spring. Buried by more calcium carbonate sediment, the algae died and rotted, leaving the cavities.

Continue up the hill on Columbia another block to 4th Avenue and turn left (north) to the Rainier Club (820 4th Avenue). Fear not, we aren't going inside "Seattle's preeminent private club," so you needn't worry about breaking the dress code. You have come to pay homage to something grander and more venerable, America's most widespread and famous building stone: limestone quarried from the Salem Formation in south-central Indiana. It caps the brick walls facing 4th Avenue. The Salem, also known as Indiana limestone, is chock-full of marine fossils, which passersby blithely ignore. Once you begin examining the stone with your hand lens, however, they probably won't ignore you, though many may shy away.

The Salem Formation was deposited in the Mississippian and Pennsylvanian periods, 359 to 299 million years ago. The Midwest was submerged beneath a shallow tropical sea. The waters were ideal for the growth of reefs, and the limestone is their remnant. With a hand lens you can see the rock consists mostly of rounded grains known as ooids (rhymes with ovoid; both words mean "egg shaped"). Ooids are sand-sized spheres of calcium carbonate deposited around a shell fragment or other nucleus, such as a sand grain. Currents sweeping across the seafloor round them. Limestone consisting mainly of ooids is called oolitic limestone or oolite (oh-uh-lite). The nearest town to the main quarries of this stone is appropriately named Oolitic, Indiana.

Mixed in with the ooids you are sure to find round disks up to 0.5 inch (13 mm) across. These are cross sections of the stalks and fans of crinoids, many-branched, calcareous, sedentary animals related to sea stars and urchins. Anchored to the seafloor with a rootlike holdfast, crinoids (also known as sea lilies) past and present rise above the seafloor on delicate stalks of layered disks. At the top of the stalk, feathery arms wave about in the current catching tidbits to eat. The stalks of ancient crinoids were typically 3 feet (1 m) tall, but one fossil giant was 130 feet (40 m) long! These animals came close to total extermination during the Permian-Triassic mass extinction, but a few species survived. Six hundred species remain today.

You may see more complexly branched and netlike fossils 0.1 to 0.25 inch (3 to 6 mm) across. These are fragments of great colonies of bryozoans (moss animals), tiny invertebrates—each called a zooid—that build soft, spongy cities on the seafloor. The colonies consist of zooids in the millions, each surrounded by an exoskeleton of calcium carbonate. The delicate colonies fall apart as zooids die, leaving behind heaps of the calcareous skeletons piled into drifts by currents and waves.

Brachiopods are less common in the Rainier Club stones. These were invertebrates with a clam-like shell at the end of a stalk. They left behind shells considerably larger than the fragmented fossils of crinoids or bryozoans. Like the bryozoans, they filtered food out of the water. Brachiopod fossil shells are typically thin, curved, broken shards when seen in cross section.

Some fossils in the Salem limestone are feces. Look for looping squiggles up to 3 inches (8 cm) long, perhaps left by a snail as it crawled across the seafloor.

Many Paleozoic limestones consist almost entirely of crinoids. These Silurian-age Moroccan fossils are on display at Western Washington University's hallway geology museum (see vignette 21).

branched fan

stacked disks

stalk

The Indiana limestone at the Rainier Club is composed entirely of fossils. The round disks are crinoid stems.

Limestone from the Salem Formation is found on buildings in virtually every state. It is relatively inexpensive due to its central location and easy to cut because of the softness of calcium carbonate. Salem limestone was used in the Empire State Building, Grand Central Station, the Lincoln Memorial, and federal buildings far and wide, including Bellingham's.

Continue north on 4th Avenue for four blocks and cross University Street. Stop and look kitty-corner downhill at the twenty-seven-story brick-clad Seattle Tower (1218 3rd Avenue). Now dwarfed by its concrete, steel, and glass neighbors, this ziggurat-style art deco building is an allegory for a mountain. The original owner, the Northern Life Insurance Company, wanted a building that reflected strength and permanence, so the building was designed to resemble nearby mountains. The top of the building steps back to form a narrow summit. The exterior bricks are dark at the bottom, becoming progressively lighter toward the "summit." Thirty-three shades of bricks were used to create this illusion. The color pattern suggests that we have figuratively climbed above tree line to the snowy summit.

At University Street turn up the hill to 5th and cross to the east side. Stop outside the Brooks Brothers store (1330 5th Avenue). The building's salt-and-pepper granitic foundation is our old friend the Index granodiorite. The creamy tan sandstone above is from the Eocene-age Carbonado Formation. Sandstone has been quarried at Wilkeson, northwest of Mount Rainier, since 1886 and is better known as Wilkeson sandstone. It is one of the most widely used Washington quarry stones. Wilkeson is distinguished by its pale color, caused by muscovite (a silvery mica); this sandstone is quite different from the darker and slightly older Chuckanut sandstone. The sediment was deposited in floodplains and deltas 45 to 40 million years ago. The sandstone is about 50 percent quartz and 35 percent feldspar crystals. The rest is mica crystals and tiny clasts of old plutonic, metamorphic, and marine sedimentary rocks. All the sediment was eroded from uplifted rocks in the Cascade Range, which was just beginning to rise. The reddish tone comes from iron oxides coating the quartz grains. Dark layers of slightly coarser, more-oxidized grains highlight bedding planes.

At Washington Federal Savings, along 5th between Union and Pike, the creamy rock is the Italian travertine from Tivoli. It has been desecrated; the voids have been filled with dark-gray caulking, perhaps in the mistaken belief that the pockets can allow water to enter the rock — under awnings, no less!

The sidewalk on the west side of 5th Avenue between Pike and Pine consists of four different urbanite paving stones mixed together.

The black ones are asphalt concrete, commonly known as asphalt or blacktop. Asphalt, or bitumen, is the residue left after gasoline, kerosene, and other useful products have been distilled from crude oil. For paving stones and roads, asphalt is mixed with sand and gravel to make asphalt concrete. Technically *concrete* refers to any composite made of a mineral aggregate, usually sand and gravel, mixed with a binder, such as asphalt, portland cement, or epoxy. Portland cement is a mix of ground and baked limestone, clay, bauxite, gypsum, and calcium sulfate. The other pavers in this sidewalk are also concrete, made of Portland cement and plenty of small pebbles in one case, fewer pebbles in the lightest-colored pavers, and mostly sand in the dark-gray ones. A few steps farther north, at the corner of 5th and Pine, three different granitic rocks make up the geometrically arranged pavers.

Cross Pine to Westlake Center (400 Pine Street), which is open seven days a week but does not permit photography inside. This building and the transit station below are a treasure trove of geology. You can spend a long time looking at all the stone in here. The floor is tiled with fossil-rich Jurassic limestone from the Comblanchien Formation, quarried from Burgundy in France. You'll have to crawl around to see fossils of bivalves, corals, and gastropods. (Tell security you lost an earring.) The 160-million-year-old Comblanchien, and other limestone formations in the Burgundy region, weather to the excellent soils that grow the famed Burgundy wines. This stone was used in the construction of the Paris Opera.

Much larger and younger fossils lurk upstairs. Take the escalator to the second floor and turn right at the top. The doorway of the first store around the corner is flanked with tan pillars, probably from the Key Largo Formation of the Florida Keys. The limestone is almost made entirely of the calcium carbonate exoskeletons of millions of tiny coral polyps that lived in reef colonies. Saw cuts on the unpolished limestone often obscure the detail, but the characteristic radiating tubes in the coral are evident; these were galleries that the polyps lived in. The Key Largo limestone is about 130,000 years old, but none of the coral species found in it are extinct. These corals are reportedly the largest fossils in Seattle building stones. A pink conglomerate, age and provenance unknown, frames the windows of this store.

Descend two levels to Metro Transit's Westlake Station foyer above the light-rail tunnel. The dimly lit corridor is a petrologic potpourri of igneous, sedimentary, and metamorphic rocks. The floor has large squares of granite dominated by huge pink orthoclase feldspar crystals. The rock with the swirled texture is Morton

A fine coral fossil in the limestone pillars upstairs in the Westlake Center, taken with a special one-time permit.

Gneiss. Smaller dark-red granite squares, of unknown provenance, set off the other stones. The pale stone on the walls and ceiling and the rich yellow rock with wood-like texture are both varieties of travertine from New Mexico, deposited at least 321,000 years ago.

The dark rock on the pillars supporting the diamond-shaped lights and on other trim is 500-million-year old Brazilian charnockite; it is easier to see this rock downstairs on the light-rail platform. Charnockite is a coarsely crystalline, nonfoliated metamorphic rock derived from plutonic rocks rich in orthopyroxene. The dark-brown and honey-colored mineral crystals are orthopyroxene. The bluish crystals are plagioclase feldspar, which flash as you change angles to look at them. This characteristic is called chatoyance (from the French *œil de chat*, for "cat's eyes"); it is caused by light reflecting off microscopic deformities in the crystals. The metamorphism of charnockite occurs at high temperatures found only in the lower crust of continents. The heat required to metamorphose the plutonic rock was probably provided by the intrusion of granitic magma.

Charnockite is interesting to petrologists because it is a sample of deep continental crust. A combination of profound uplift, erosion, and time are required to bring it to the surface from so deep inside ancient continents. Charnockite is found all over the world, most often in deeply eroded Precambrian rock. There are many varieties used as building stones with a profusion of trade names. It is most

On the walls beside the light-rail tracks are panels of dark charnockite. The honey-yellow and creamy travertines are from New Mexico. The source of the dark-red granitic rock is unknown.

often called some variety of green granite or Ubatuba granite, after the Brazilian city near some of the earliest quarries – and the source for the Westlake Center rock. Charnockite is now quarried in many places, including India, Labrador, Norway, South Africa, and China.

Exit the transit station to Pine and 6th. The walls at the exit are sheathed with granite bearing red garnet crystals. Walk east on Pine Street to the Grand Hyatt Hotel at 7th and Pine (721 Pine); it's open all week. Enter the building using the Pine Street entrance. Go straight through the lobby and descend the steps toward the meeting rooms.

The rock on the floor is fossil-filled Late Jurassic limestone from central Germany's Treuchtlingen Formation, deposited in the Tethys Ocean about 155 million years ago. The Portoro limestone at the Exchange Building was also deposited in this ocean, but millions of years before. Most of the fossils are cup-shaped sponges. The real prizes are the coiled ammonite fossils. These were free-swimming, shelled, and tentacled cephalopod mollusks that went extinct in the same catastrophe that killed the dinosaurs 65 million years ago. Nautiluses are their modern descendants. A fine 5-inch-long (13 cm)

Calcite crystals fill part of this ammonite fossil on the steps down to the amphitheater. Look closely to see the corrugated margin of the shell.

The translucent onyx marble in the Grand Hyatt is only 0.4 inch (1 cm) thick.

ammonite is on the right side of the steps down to the Eliza Anderson Amphitheater. Some of the ammonites are partially filled with white calcite crystals. Several still show the individual chambers in the shells. As the animal grew, it occupied only the portion of the shell near the opening; the rest was filled with water. This provided the animal nearly neutral density, helping the ammonite to float in a steady position instead of sinking.

All along you have probably been casting glances at the lustrous backlit screen at the end of the hall that resembles an enlarged photo of pizza crust or the martian surface. The panels are actually 0.4-inch-thick (1 cm) slabs of onyx marble (yes, this is rock!) from the Eskisehir region in northwestern Turkey. Like any rock, this onyx marble is translucent when sliced thinly enough. Geologists routinely grind and polish ultrathin sections of rock that are only 0.001 inch (30 microns or 0.03 mm) thick for study. The technology and skill necessary to cut and polish the Hyatt's relatively thick rock panels and keep them intact during installation is impressive. This is the finale of your tour, the ultimate in quarried and polished rock, on display as an art piece with a view through the entire rock.

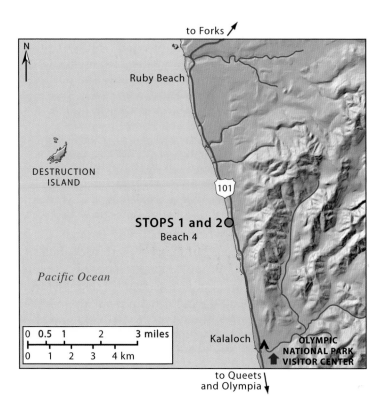

to Forks ↗

N

Ruby Beach

DESTRUCTION
ISLAND

101

STOPS 1 and 2○
Beach 4

Pacific Ocean

0 0.5 1   2   3 miles
0   1   2   3   4 km

Kalaloch

OLYMPIC
NATIONAL PARK
VISITOR CENTER

to Queets
and Olympia ↓

GETTING THERE: Beach 4 is on the wild Pacific shore of Olympic National Park, 2.7 miles (4.3 km) north of the popular Kalaloch Campground on US 101. From the south, the beach is 73 miles (117 km) north from Hoquiam. From the north, it's 88 miles (142 km) from Port Angeles via Forks. The well-signed turn into the paved Beach 4 parking lot is halfway between mileposts 160 and 161.

From the parking lot, a trail descends an easy 100 feet (30 m) along a small stream, reaching the beach and stop 1 in 200 yards (180 m). Unless the tide is high, walk another 200 yards (180 m) or so up the beach to the north to stop 2. It is possible to walk farther along this gorgeous beach for miles to the south or north; popular Ruby Beach is about 4 miles (6 km) to the north. Be careful you don't get trapped against the coastal bluff by high tide and surf. Always consult a tide chart before setting out.

15

THE FOLDED ROCKS AT BEACH 4

From the parking lot the main trail descends to the beach. But first, follow the right trail fork for 50 yards (50 m) to a fine overlook above the beach. This will give you a chance to see the tide level on the beach. Flat-topped Destruction Island lies 3.5 miles (5.6 km) to the northeast. Once part of the coast, it has been isolated by coastal subsidence and erosion.

Take the main trail downhill to stop 1, the outcrop at the foot of the trail. The steeply tilted sedimentary rock layers rising out of the sandy beach were originally deposited as close-set, horizontally stratified layers of sediment on the seafloor. The grains in the sediment were compressed and cemented together, or lithified, by pressure within the crust, becoming rock. Later, the layers were compressed, folded, and in many cases fractured to bits as they were subducted under North America. The rocks were then uplifted to sea level, where waves planed them off, truncating the beds. The eroded rocks were later covered by younger flat-lying sedimentary beds, which have yet to be lithified. It's what we don't see here that is the focus, represented by that angular contact between the dipping rocks and the gravel.

An unconformity is a time gap in the geologic record, where rocks of significantly different ages are in contact with each other. The contact between the vertical and the horizontal layers at stop 1 is an angular unconformity. This means that two rock formations, each with layered beds dipping at different angles, are separated by an erosional surface truncating the lower rocks. For this to occur, compression or faulting must first fold or otherwise tilt the lower rocks. Then, sufficient time must pass to allow erosion to strip off any younger layers that may have lain higher in the geologic sequence before beveling off the folded, tilted layers. And then younger sediment is deposited on the erosion surface. An angular unconformity of necessity represents a great gap in time. Consequently, the single small rock outcrop at Beach 4 encapsulates a lot of geology and evokes a story from the dawn of modern geology.

Ocean view from the bluff near the parking lot. The rocky islet just off Starfish Point is a wave-eroded sea stack. Low-lying, flat-topped Destruction Island barely rises above the ocean horizon, upper left, 3.5 miles (5.6 km) to the northwest.

The Olympic Subduction Complex at stop 1 is overturned so that the youngest beds are toward the left of the photo. The contact between these rocks and the overlying Browns Point Formation gravel is an angular unconformity.

angular unconformity: younger sediments rest upon an older surface of tilted or folded rocks. An obvious erosion surface.

nonconformity: an unconformity between stratified rocks above and unstratified igneous or metamorphic rocks below. An obvious erosion surface.

schist   granite

disconformity: an unconformity between beds that are parallel. An erosion surface is not obvious.

marine rocks

river deposits

The types of unconformities (red line).

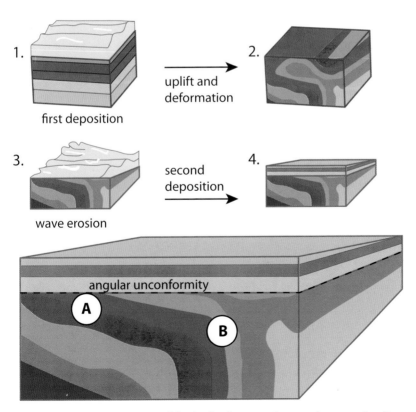

1. first deposition

uplift and deformation

2.

3. wave erosion

second deposition

4.

angular unconformity

A

B

Long periods of time are required for the development of an angular unconformity. Some beds are overturned (A), while others (B) are vertical.

James Hutton (1726–1797), a Scottish naturalist, physician, and experimental farmer, is known as the father of geology. Hutton had tremendous insight into the age of the Earth in 1788, when he saw an angular unconformity at Siccar Point on the North Sea coast of Berwickshire in southeastern Scotland. Siccar Point is one of the most famous geological localities in the world, and Beach 4 is a convenient local substitute for it.

Hutton had been looking for a geologic site that could demonstrate his idea that Earth was far older than that accepted by prevailing views. Europeans of the day largely agreed with the interpretation of Archbishop Ussher of Ireland, who, in 1650, used biblical genealogies to calculate that Earth was created late in the afternoon of Sunday, October 23, 4004 BCE. Hutton famously said that "the present is the key to the past," meaning that Earth processes had proceeded at the same rate in the past as they do today, a geologic principal today known as uniformitarianism. He maintained that rocks formed at imperceptibly slow rates in the past and were worn away by erosion at a rate imperceptible to humans, just as they are today. The resulting sediment was again buried, lithified to form new rocks at some (in Hutton's time) unknowable depth, and then uplifted and exposed to a new cycle of erosion. At Siccar Point, Hutton found what he'd been looking for: steeply tilted layers of dark-gray sandstone rising out of the sea, truncated at a sharp angle, and covered with nearly flat-lying beds of red sandstone.

Hutton recognized that the vertical gray beds were at one time horizontal seafloor sediment that had lithified ever so slowly to sandstone. They were tipped nearly vertical, raised above sea level, planed off by erosion, then again submerged beneath the sea to be covered with the sediment that would become the red sandstone. Hutton figured that far more than 5,792 years had elapsed between Ussher's creation date and the year Hutton recognized the unconformity, and that the unconformity itself represented a vast gulf of time. He stated this in his most famous quote: "We find no vestige of a beginning, no prospect of an end." We now know that Hutton's unconformity represents a time gap of 80 million years, separating 345-million-year-old Old Red Sandstone from 425-million-year-old Silurian sandstone.

In contrast to Hutton, the influential German mineralogist Abraham Werner (1749–1817) maintained that Noah's Flood was the source of all sediment and sedimentary rocks. This concept is called catastrophism, in contrast to Hutton's uniformitarianism. Practitioners of the new science of geology (the term had only been coined in 1778) were divided between the two schools of thought. Hutton

The unconformity at Siccar Point is perhaps the world's most famous geologic site. An angular unconformity (yellow line) separates a 3-foot-thick (1 m), reddish, nearly horizontal sandstone from vertical gray sandstone below. White lines highlight the dip of the beds. —Courtesy of Dr. Clifford E. Ford, School of GeoSciences, University of Edinburgh

published his ideas in *Theory of the Earth* (1785). Unfortunately, his writing was obtuse in the extreme, and the book remains a tough read. It was left to others to restate his ideas, which were gradually accepted. Uniformitarianism is now a bedrock tenet of geology, taught in all beginning geology courses.

You may find references that call these sedimentary rocks along the Olympic coast the Hoh Formation. Geologists use the term formation for rocks that are contemporaneous, distinctive from others in the area, related in their depositional environments, and extensive enough to be mapped on the surface or traceable in the subsurface. The sedimentary units along the Olympic Coast are a mishmash of different-looking rocks of widely varying ages and degrees of deformation. Some are highly contorted and fragmented; others, like the rocks at Beach 4, are folded but not much contorted. None of the Hoh rocks are laterally continuous enough to be mapped as distinct units over any distance. Consequently, the sedimentary rocks up and down the Olympic National Park coastal strip cannot be considered a formation. Rather, they have been placed in two separate terranes, the Ozette and the Hoh, separated from each other by faults. The Hoh and Ozette Terranes have been lumped into a catchall group, the Olympic Subduction Complex. This name recognizes a common historic thread: all are marine rocks deposited somewhere off the coast, and all were subducted beneath North America to some degree.

At Beach 4 the tilted rocks, part of the Olympic Subduction Complex, are Miocene-age turbidites deposited between 23 and 16 million years ago; they are composed of sandstone and mudstone. Turbidites have a distinctive, repetitive pattern of beds that result from turbidity currents: flows of suspended mud, sand, and even gravel that travel across the seafloor. These flows may be derived from rivers or debris flows entering the sea, or submarine landslides on the continental shelf. Turbidity currents can flow at high velocities and for great distances. In 1929 an earthquake off the coast of Nova Scotia collapsed a submarine slope on the Grand Banks. The suspended debris raced across the floor of the Atlantic Ocean for at least 180 miles (290 km) at speeds of 40 miles (65 km) per hour, breaking transatlantic cables and interrupting service.

When turbidity currents first begin to slow down, coarse-grained sand and gravel particles settle onto the bottom quickly. Fine-grained mud remains suspended in the water, settling much more slowly over the coarser sediment. This type of layering—coarser particles at the base and finer particles at the top—is said to be graded in a fining-upward sequence. Repeated couplets of coarse and fine sedimentary layers, indicating multiple individual events, are a hallmark of turbidites.

At stop 1 the graded turbidites form couplets of buff sandstone and darker-gray shale. Find a place where a buff layer is sandwiched between two darker layers. If the contact, or boundary, between the sandstone and darker shale is very sharp, you are looking at shale from one turbidite and sandstone from a different one. If the boundary is not sharp, the dark and buff layers are part of the same turbidite. Most of the beds along this beach are upside-down, so older turbidites lie above younger ones. The tipped rocks record scores of turbidity currents. Their sheer number indicates that unstable slopes persisted for a long time off the continental coast where this sediment was deposited.

How did the formerly submarine Olympic Subduction Complex rocks get here on the shoreline? What tilted them? Was the sediment originally deposited on the seafloor right here, and then raised to

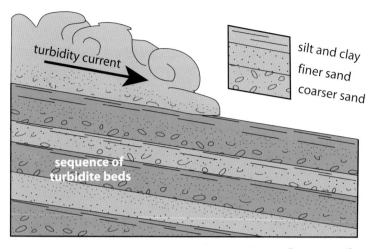

Turbidites are deposited when sediment mixed with water flows across the ocean bottom. Coarse-grained clasts settle out first, to be covered by finer-grained particles settling out of the suspended cloud.

just above sea level? Turbidites are typically deposited seaward of continental shelves, in deep ocean basins. The sparse fossil record preserved in these rocks shows that sedimentation occurred in water depths exceeding 6,600 feet (2,000 m). A blanket of sediment about 1.5 miles (2.4 km) thick lies on top of the eastward-migrating Juan de Fuca Plate. In a process that continues today, this thick sedimentary blanket is crushed against the North American Plate at the plate boundary, about 80 miles (130 km) off the Olympic coast, and is scraped off to form what's called an accretionary wedge along the outer edge of North America. The accretion of more seafloor sediment to the western margin of the wedge shoves the previously accreted sediment farther east. Burial beneath the sediment blanket and pressure from the plate collision turns the sediment to rock. The lateral pressure crumples the lithifying sediment and fractures it as well, rumpling the layers like a throw rug your dog made a high-speed turn on.

A small portion of the chaotic jumble of the Miocene part of this wedge is what we see at Beach 4. These tilted rocks have migrated laterally 80 miles (130 km) or so from the subduction margin off the coast but were not subducted. However, they were deformed enough to be overturned in places, and they were gradually uplifted to just above sea level, where they were subjected to wave erosion at the shoreline. A calculation based on the area of the wedge and

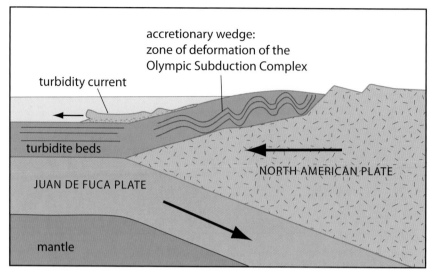

Deposition and deformation of the Olympic Subduction Complex. Turbidites piggybacked on the Juan de Fuca Plate are crushed against the backstop of the North American Plate.

the steady, uniform movement of the two plates indicates it took around 22 million years for the rocks to migrate to the coast. This estimate independently agrees with the age of these rocks. The youngest tilted rocks in the stack at stop 1 are nearest the sea.

The turbidites are pitted with shallow holes a little above the high tide line. These are rarely more than 1 inch (2.5 cm) or so deep and 1 to 2 inches (2.5 to 5 cm) across. They are the remains of once-deeper burrows of the rock-boring clam *Penitella penita*, also known as the piddock clam. When these clams are young and small, they burrow into soft rocks rather than mud on the seafloor, as most clams do. A piddock clam attaches its fleshy foot to rock and pivots the hard edges of its shell back and forth to grind away the rock grain by grain. The resulting pit becomes deeper; as the clam grows, the cavity widens within the rock, and the clam becomes trapped in its own hole. The clam, which may grow to 3 inches (7.5 cm) in length, may end up at the bottom of a pit 6 inches (15 cm) deep. These clams either live in the lower part of the intertidal zone or entirely subtidally. These borings above sea level are further evidence for the uplift and erosion of these rocks.

The turbidites were truncated by waves around 122,000 years ago. At Beach 4 this wave-cut bench is about 9 feet (3 m) above today's mean sea level. The bench can be traced for 50 miles (80 km)

Piddock clam burrows have been uplifted above sea level. No shells remain in these eroded holes.

Piddock clams in sandstone.

along the coast. In places it dips below sea level; south of Kalaloch the bench rises as much as 160 feet (50 m) above sea level. The sediment above the truncated top of the tilted turbidites, and extending to the top of the bluff, is crudely stratified sand, gravel, and peat of the Pleistocene Browns Point Formation. Streams flowing from large glaciers in the Olympic Mountains during Pleistocene ice advances deposited this outwash. Radiocarbon ages collected from the Browns Point Formation range from more than 47,000 years old at the bottom to about 16,700 years near the top. The contact between the turbidites and the Browns Point Formation is an angular unconformity. The unconformity represents a little less than 16 million years of missing time, the difference between the minimum age of the Olympic Subduction Complex (16 million years) and the maximum age of the Browns Point Formation (47,000 years). This is small potatoes compared to Siccar Point's 80 million years,

The fining-upward gradation of grains in turbidites shows that these beds have been overturned. The dashed line indicates the transition from coarse to fine grains. The contacts at the top and bottom of a single turbidite sequence are knife sharp.

or the Great Unconformity at Grand Canyon, which represents 250 million years or more of missing time, but it is much easier to visit!

Walk north along the beach. You'll notice that the dip of the turbidites changes. In places they are nearly horizontal but entirely upside-down, having been overturned during folding. This is not immediately obvious; however, through careful examination of the changes in grain size in individual layers, and by identifying the usually distinct, planar contacts between individual turbidites, a geologist can usually figure out which way was originally up and younger in a sequence, even if the whole sequence is upside-down.

At stop 2, about 200 yards (180 m) north of the trail at the base of the bluff, a wonderful sequence of chevron-shaped folds is exposed. The amount of exposed rock depends on the depth of sand against the bluff, which varies from 3 to 6 feet (1 to 2 m). The sharp line of a thrust fault separates the folded turbidites from south-dipping layers beneath, also turbidites. Faulting occurred as these rocks were scraped off the subducting Juan de Fuca Plate. The rocks above the fault were displaced sideways an unknown distance relative to the

The gorgeous, textbook chevron folds at stop 2 are offset by a thrust fault above a different set of dipping turbidites. —Courtesy of Gene Kiver

dipping layers. Sharp, small-scale folds like these are a rarity in western Washington and alone are worth the pilgrimage.

A short distance before the rocky point at the north end of the beach (locally known as Starfish Point) you can see a different breed of unconformity. In places the turbidite beds are horizontal, whether flipped upside-down or not, and are overlain by the Browns Point Formation sand and gravel, also horizontal. However, there is still a time gap, called a disconformity in this case. This term is applied when successive geologic layers are parallel but of different ages. A geologist might overlook a disconformity between two similar-looking rock formations until detailed observations elsewhere reveal that erosion made a time gap between them. Here, the unconformity is obvious because the turbidites and unlithified Browns Point sand and gravel look nothing alike.

The disconformity between the horizontal but overturned reddish turbidites and overlying horizontal, darker Browns Point gravel is obvious.

Starfish Point is less than 400 yards (360 m) from stop 1. Beyond the point, the beach continues north about 4 miles (6 km) to Ruby Beach, passing one or two other access trails along the way. If you choose to continue, check a tide table first to reduce your chance of getting caught by an incoming high tide.

The sea stack at Starfish Point consists of turbidites that are relatively erosion resistant. It is a remnant of the coastline when it was 300 yards (270 m) or so farther west than today. The implacable ocean is eroding the point to fragments; eventually the sea stack will disappear altogether. To the northwest, flat-topped Destruction Island is a remnant of the coast when it was even farther west, during the great lowering of global sea level during the Pleistocene. The same wave-cut unconformity seen at stop 1 forms the island's surface. Destruction is the first island of any size north of the Farallon Islands (west of San Francisco) and the largest off the Washington and Oregon coasts. Sea stacks become increasingly common farther north along the wilderness coast of Olympic National Park; some of them are well over 200 feet (60 m) high. All will eventually be worn away, replaced by new ones as the hungry ocean eats into rock of the coast.

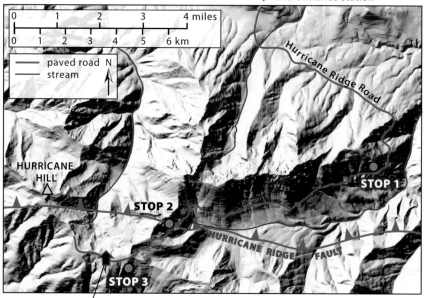

to Port Angeles
and entrance station

HURRICANE RIDGE VISITOR CENTER

GETTING THERE: Hurricane Ridge is reached by way of the winding Hurricane Ridge Road in Olympic National Park. From US 101 in Port Angeles, follow prominent signs leading south on Race Street. The Olympic National Park Visitor Center is at the edge of town and is a good place to get oriented. Immediately south of the visitor center, go right on Hurricane Ridge Road at the junction with Mount Angeles Road, from which the road climbs higher and higher into the park. It is important to set your odometer to 0 at the entrance station, about 5 miles (8 km) from the road junction. Stop 1 is at mile 5.7 (9.2 km). Stop 2 is 100 yards (90 m) beyond milepost 16, at mile 10.9 (17.5 km) on your odometer. There are paved pullouts on both sides of the road at both stops. Stop 3 is the popular Hurricane Ridge Visitor Center at mile 12.6 (20.3 km). Plan this trip to catch the morning sun at the first two stops. Since the stops are along a popular road in the national park, you might want to bring along a bright jacket or shirt to wear for added visibility. You are in a national park, so absolutely no hammering or collecting is permitted.

16

Seafloor in the Sky

HURRICANE RIDGE

The Olympic Peninsula is dominated by the Olympic Mountains, with their apex at icy Mount Olympus (7,979 feet, 2,432 m). Two sets of unrelated rocks, one wrapping around the other like a taco, compose the broad story of Olympic Mountains geology. The rocks in the core of the mountains and extending west to the Pacific shore are the filling. These are mostly Paleocene- to Miocene-age marine sandstone and shale, tortured to various degrees by subduction and accretion. They are actually a set of four roughly similar terranes, collectively known as the Olympic Subduction Complex (vignette 14), but appear in some publications as the Core Domain, Olympic Structural Complex, or Olympic Core Rocks.

The Crescent Terrane (the Peripheral Domain of older references) wraps around the Olympic Subduction Complex on the north, east, and south and is the tortilla in our analogy. A terrane is a grouping of rock that is bounded on all sides by faults and has a distinctly different geologic history from adjacent rock units. Terranes move into position adjacent to each other through tectonic activity. The Crescent Terrane has a similar age span as the Olympic Subduction Complex. It consists of lava flows of the Eocene-age Crescent Formation, largely submarine, although some were erupted above sea level. Radioisotope dates for the basalts range from 58 to 31 million years old. Geologists have calculated that the volume of basalt in the Crescent Terrane is upwards of 24,000 cubic miles (100,000 km³). A section of basalt in the Dosewallips River valley, on the east side of the peninsula, is 10 miles (16 km) thick. The Crescent basalt may be the thickest section of basalt in the world! The Blue Mountain unit, which lies below or interfingers with the basalt, is among the sedimentary layers composing the terrane. The youngest overlying sedimentary rocks are 23 to 20 million years old.

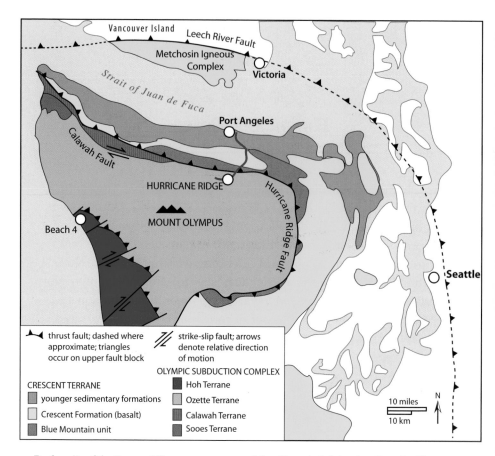

Rock units of the Crescent Terrane wrap around the Olympic Subduction Complex like a giant horseshoe. On Vancouver Island, the Crescent Terrane is called the Metchosin Igneous Complex. (Modified from Babcock et al. 1992.)

The Hurricane Ridge and Calawah Faults separate the Crescent Terrane from the Olympic Subduction Complex. The complex has been shoved obliquely beneath the Crescent Terrane along these faults. The Leech River Fault, another thrust fault, separates the Crescent Terrane from the rocks above it. This faulted contact is only well exposed along the southern edge of Vancouver Island. Geologists assume this fault parallels the Hurricane Ridge Fault, curving southward under Puget Sound.

What complex geology! One set of rocks is stuffed beneath another set, although both are about the same age. The fault that separates them is a long, sharply curving horseshoe. The rocks were deposited or erupted on the seafloor but are now a wild mountain range.

Here's the kicker: the Olympic Subduction Complex and the Crescent Terrane were deposited on different tectonic plates. It's all a fine mess for geologists to puzzle over.

There are no roads into the interior of the range, and only logging roads nibble at the margins. Beyond the roads, junglelike forest and brush cloaks the rock, and then there is the famously prodigious rainfall, which, with periodic glacial advances, has eroded steep valleys and hillsides. These all combine to make fieldwork in the Olympics a strenuous adventure, every time.

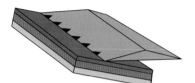

1. Shove the Olympic Subduction Complex under the Crescent Terrane rocks at the leading edge of North America.

2. Compress the whole region northward against the thick crustal "backstop" of British Columbia's interior.

3. Uplift the rocks as erosion grinds away for 14 million years.

How to make an Olympic horseshoe.

To understand the origin of the Crescent Terrane, with its thick basalt, we must consider where such submarine basalts originate. If we assume the basalts erupted onto an oceanic plate, there are a couple of tectonic settings where this might happen. The first is the rift separating spreading ocean plates. As an example, basalt lavas erupt every now and then at the spreading center off the coast of Washington, where the small Juan de Fuca Plate is separating from the huge Pacific Plate.

The other setting is called a hot spot, a blowtorch of magma rising out of the mantle. In this case, huge masses of basalt build volcanic island chains such as the Hawaiian Islands. The monstrous pile of basalt that makes up the Big Island in Hawaii is capped by Mauna Kea, 13,796 feet (4,205 m) above the sea. However, the mountain's base is about 18,000 feet (5,500 m) below sea level. So, at around 31,800 feet (9,700 m), it is taller than Mount Everest, a mere 29,035 feet (8,850 m) above sea level. Hot spots can also spawn seamounts, large masses of basalt that do not protrude above the waves. Close to home, the Cobb Seamount is a submerged Oligocene basalt volcano about 300 miles (480 km) west of Grays Harbor. It rises to within 112 feet (34 m) of the surface.

Now let's consider how the two distinct packages of rocks of the Olympic Peninsula came to be nestled together. There are two contrasting interpretations. The theories differ mainly around the origins of the Crescent Terrane. The first posits that thick sediment and volcanic masses, including seamounts and extensive sheets of ocean floor lava flows, were scraped off an oceanic plate and accreted to the leading edge of the North American Plate as the oceanic plate was subducted. In this theory, the Crescent Formation basalt originated as seamounts erupted onto the oceanic plate, and the Crescent Terrane sedimentary rocks began as sediment on the seafloor

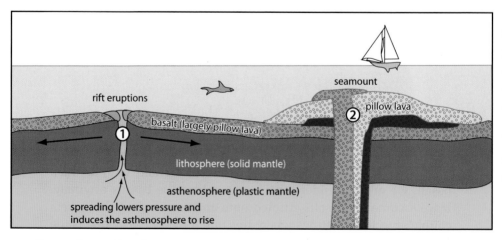

Large volumes of basalt can accumulate on an oceanic plate in two tectonic environments: at spreading centers (1), where two oceanic plates are separating, and at hot spots (2), where magma piles up on the surface as a seamount. Big seamounts may breach the ocean surface to form island chains, such as the Hawaiian Islands.

around and on the seamounts. The thick mass of the Crescent Terrane jammed the subduction zone, so that subduction jumped to the western margin of the Crescent Formation. The Olympic Subduction Complex rocks originated as seafloor sediment farther offshore. When they reached the subduction zone, they were able to slide under the Crescent Terrane but were soon scraped off and accreted, too. The Hurricane Ridge Fault developed as a thrust fault at the boundary of these two packages of rock. The whole shebang was then uplifted, folded, and tipped downward to the east.

The second, newer theory hinges on a different explanation for the origin of the Crescent Terrane. When the Crescent Formation basalt was erupted, the North American Plate was mostly riding over the oceanic Farallon Plate. This huge plate has now largely disappeared beneath North America; the relatively tiny Juan de Fuca, Explorer, and Gorda Plates are the remaining scraps. However, there was another oceanic plate, the Kula, in the northeast corner of the Pacific Ocean, which moved northward relative to North America. It has been completely subducted at the Aleutian subduction zone.

This theory proposes that during the Eocene, at the margin of the North American Plate, where the Olympic Mountains would eventually rise, the continent was overriding the spreading center separating the Farallon and Kula Plates. Even though the rift dove beneath the margin of North America, it was still widening. Normally at a spreading center magma rises and fills the gap with rapidly solidified lava. However, some researchers believe the growing gap, called a slab window, provided a route for huge volumes of basaltic magma to rise out of the mantle through the thin crust at the continent's leading edge, filling the slab window and erupting onto the shallow seafloor. This lava became the Crescent Formation basalt, and the seafloor sediment of the continental edge became the sedimentary formations of the Crescent Terrane above and below the basalt. According to this interpretation, during the Eocene there were large, overlapping volcanic islands in the shallow sea just off the coast, and the Crescent Terrane is actually part of the overlying continental plate rather than comprising the oceanic plate seamounts and seafloor sediment suggested by the other interpretation.

Either way, continued subduction and accretion of continental shelf sediments (see vignette 15) is driving the Olympic Subduction Complex beneath the Crescent Terrane rocks and is contributing to the steady, slow rise of the Olympic Mountains. The North American Plate is being deformed upward as it and the Juan de Fuca Plate shove against each other (see the introduction and vignette 4).

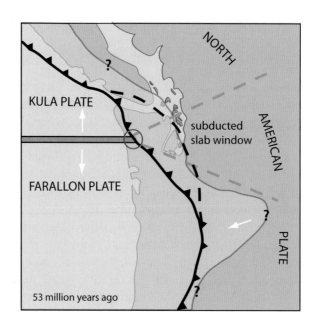

NORTH

KULA PLATE

AMERICAN

subducted
slab window

FARALLON PLATE

PLATE

53 million years ago

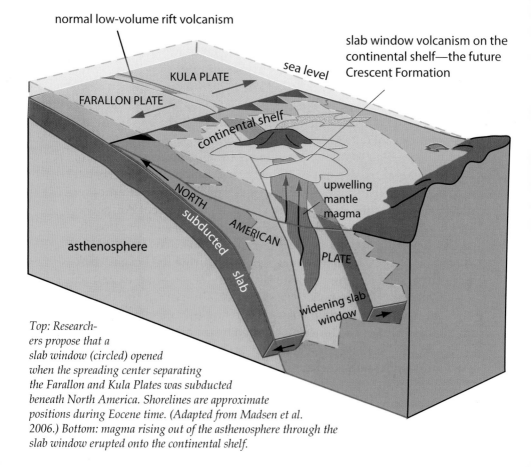

normal low-volume rift volcanism

slab window volcanism on the
continental shelf—the future
Crescent Formation

sea level

KULA PLATE

FARALLON PLATE

continental shelf

NORTH

subducted
slab

AMERICAN

upwelling
mantle
magma

asthenosphere

PLATE

widening slab
window

*Top: Research-
ers propose that a
slab window (circled) opened
when the spreading center separating
the Farallon and Kula Plates was subducted
beneath North America. Shorelines are approximate
positions during Eocene time. (Adapted from Madsen et al.
2006.) Bottom: magma rising out of the asthenosphere through the
slab window erupted onto the continental shelf.*

Several lines of evidence support the slab window hypothesis. The presence of continental shelf sedimentary rocks (the Blue Mountain unit) interbedded with the Crescent Formation basalt argues for a shallow, nearshore marine environment, not the greater depths typical of an oceanic plate hot spot. So does the basalt's chemistry; it more closely resembles that of lava erupted at oceanic spreading centers, not hot spots. The first interpretation is relatively simple to understand and fits most people's notions of plate tectonics. You are still likely to encounter it in popular interpretations and park service publications. Some recent scientific papers assume that the Crescent Terrane was accreted, though many geologists rally behind the slab window interpretation.

Why the horseshoe or taco-like structure? The Olympic Subduction Complex and the Crescent Formation rocks were together squeezed into an arch on a huge scale and stuffed into a corner of the older, thicker accreted terranes of British Columbia and western Washington (see introduction). Through the Eocene, the Crescent Terrane rocks were driven by the forces of accretion, ever so slowly slipping under older terranes of the mainland along the Leech River Fault. The oldest unfolded, undeformed rocks in the Crescent Terrane are 14.7 million years old, so we know that deformation had ended by then.

Now that you're dizzy with tectonic scenarios, head to stop 1, a tall exposure of Crescent Formation basalt on the west side of the road that erupted about 55 million years ago. The pullout is just around a curve to the right, at a small gully with trees. If you miss it, there are other pullouts very close by, but on the other side of the curvy road.

The bulbous shapes are called lava pillows. Each pillow is a separate, short-lived lobe of lava, lasting a few minutes within a longer-lived eruption. Each extends back into the outcrop. The pillowed flow accumulated over a period of a few days to perhaps months. During the eruption, separate lobes flowed alongside or on top of earlier pillows. Stop 1 is a cross section, exposed by millions of years of erosion. Cape Disappointment (vignette 3) is another great place to see pillow lava in the Crescent Formation.

Pillows form when the outer surface of lava is quickly chilled by seawater. Think of each of these pillows as a finger of incandescent lava, 1 to 2 feet (0.3 to 0.6 m) across, flowing toward you. Advance is herky-jerky. The leading tip of each lava lobe advances, akin to toothpaste squeezed from a tube. The surface of the lava is quickly quenched by seawater, forming a glassy skin, but the internal pressure from the ongoing lava extrusion at the vent almost immediately

The 60-foot (18 m) wall of pillow basalt at stop 1.

causes the hardened rind to split open. The interior lava then spurts out through the crack in another short advance, until its outer surface solidifies. The process continues until there is insufficient lava pressure to break the hard rind; this is either because the lava finds a more efficient finger to pass through, or because the eruption wanes. Spaces between pillows are often filled with glassy pillow breccia, fragments of basalt shattered during rapid quenching. If the lava is advancing down a steep underwater slope, or at a rapid pace, the lava may be mostly breccia.

The rock has a greenish cast to it, which is principally due to the minerals chlorite and epidote. Vesicles, or gas bubbles, left behind in the lava are mostly filled with white secondary minerals; most of these are in the zeolite family. White calcite forms patches and fills cracks in the lava. All these secondary minerals formed during low-grade hydrothermal metamorphism, as hot water circulated through the thickening lava pile.

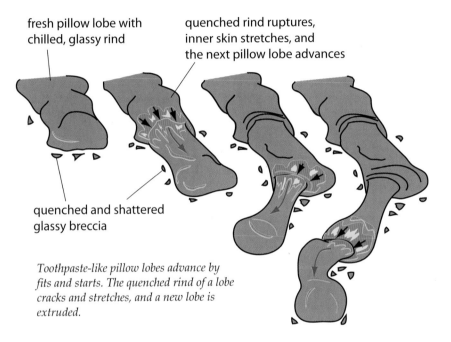

fresh pillow lobe with chilled, glassy rind

quenched rind ruptures, inner skin stretches, and the next pillow lobe advances

quenched and shattered glassy breccia

Toothpaste-like pillow lobes advance by fits and starts. The quenched rind of a lobe cracks and stretches, and a new lobe is extruded.

If you observe the pillows up close, cross sections may reveal radiating fractures inside of them. These developed once the lava lobe halted and the rock cooled from the outside in. Contraction from cooling is taken up by fractures, the same way that columns form in lava flows on the ground (see vignettes 2, 5, and 22). Some pillows are really elongated, rounded tubes; some changed direction as they grew outward. A few of the pillows can be used to show which way was down when they erupted. When a pillow lobe advances over the top of earlier pillows, the bottom of the advancing lobe fills the gap between the rounded crests to make a V-shaped lower surface. Finding one of these is among the holy grails of pillow enthusiasts since it reveals the original orientation of the stack. The pile of lava pillows at stop 1 is not upright. The whole stack dips to the north, away from the road at about 50°. Tectonic activity tilted it sometime after the eruption.

This heart-shaped pillow has a V-shaped tail (next to the knife) where lava sagged downward between two earlier pillows. Pillow tails indicate the original downward orientation of the whole stack.

Cooling fractures radiate inward in this cross section of a 2-foot-wide (0.6 m) pillow.

About 0.2 mile (0.3 km) back down the road, to the east, is a sharp, dipping contact of pillow basalt over shattered breccia. The lava initially shattered to form a layer of pillow breccia, which was then overridden by pillows later in the eruption.

When you have had your fill of pillows, head up the road to stop 2. Three shallow gullies incise the low roadcut. You are standing on a terrane boundary. The gullies have eroded along parallel strands of the east-west Hurricane Ridge Fault separating the Crescent Terrane from the Ozette Terrane of the Olympic Subduction Complex. During accretion, the subduction complex terranes were thrust under the Crescent Terrane. The Hurricane Ridge Fault is not a simple structure; it is really a number of individual, closely spaced faults, each accommodating some of the movement between the terranes. The two faults just north of the milepost cut through the base of the Crescent Terrane; the third fault, at the milepost, actually separates the Crescent and Ozette Terranes. Similar faults, farther to the left but not evident along the road, cut through the uppermost rocks of the Olympic Subduction Complex. The rocks between the gullies have been rotated and shattered by fault movement. The beds on either side of the hair-thin faults running up the centers of the gullies have different orientations.

The Crescent Terrane at stop 2 is represented by the Blue Mountain unit, steeply dipping, rhythmically bedded sandstone and slate eroded from mountains of North America and deposited on the shallow continental shelf around 56 million years ago. These turbidites (see vignette 15) carpeted the shallow seafloor before the Crescent Formation basalt lava buried them. The sharp chevron folds developed as the Ozette Terrane was accreted to North America, deforming rocks of the Crescent Terrane in the process.

The Ozette Terrane of the Olympic Subduction Complex lies on the south, or left, side of the fault at the milepost 16 marker. It consists of folded, faulted, and, in places, shattered black slate. The slate began as fine-grained sediment deposited on the floor of the deep Pacific, 100 miles (160 km) or perhaps much more off the coast of North America. The sediment first lithified to shale as a deep blanket of sediment accumulated on the seafloor. Although sedimentary structures such as bedding are recognizable, the rock was metamorphosed to slate and, in places, phyllite as it was subducted beneath the Crescent Terrane and accreted to the underside of the continental plate. The rocks of both terranes, and the faults within them, were uplifted, folded, tilted, and eroded. Quite a history!

Continue up the road to stop 3. There are great views into the interior of the Olympics from the patio outside the Hurricane Ridge

Sharp folds in Crescent Terrane sandstone lie to the right of the fault (red dashed line) in the gully at milepost 16 (lower left). Dark slate of the Ozette Terrane lies on the opposite side.

A strand of the Hurricane Ridge Fault runs up the outcrop (center of the photo; see knife) at milepost 16, separating Crescent Terrane rock (right) from dark slate of the Ozette Terrane (left). Fault movement has shattered and sheared both.

Visitor Center. The glaciated Bailey Range, 16 miles (26 km) to the southwest, is dominated by Mount Olympus itself, which is folded, broken, and faulted Ozette Terrane slate with some surviving sandstone. These rocks are easily attacked by erosion, and their steeply dipping beds form tottering pinnacles. Inside the visitor center is a geology display that interprets the entire Olympic Peninsula as a mass of crust too big to subduct beneath North America, so it was accreted en masse to the continental margin—the earlier interpretation of this region.

The high central peaks of the Olympic Mountains seen from stop 3.

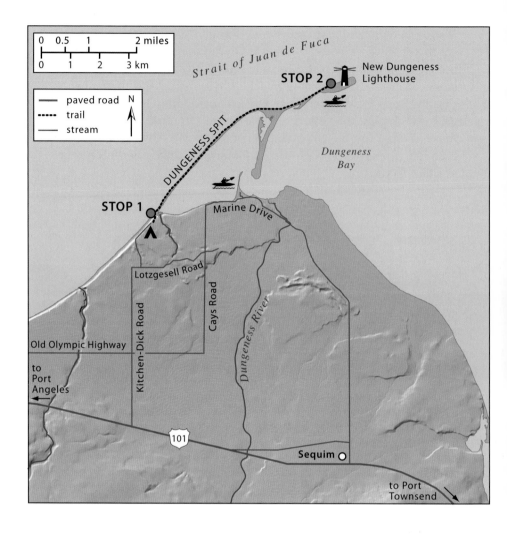

GETTING THERE: At milepost 260 on US 101, about 3 miles (5 km) west of Sequim or about 10 miles (16 km) east of Port Angeles, turn north on Kitchen-Dick Road. This intersection is signed to Dungeness County Park and Dungeness Recreation Area. Kitchen-Dick Road makes a sharp right 3.2 miles (5.1 km) from the highway and becomes Lotzgesell Road. After 0.2 mile (0.3 km) turn left on Voice of America Road and head into the county park. About 0.5 mile (0.8 km) in there are viewpoints overlooking bluffs above the Strait of Juan de Fuca. In decent weather, views extend to Vancouver Island and west to Port Angeles. Continue through the campground another 0.5 mile (0.8 km) to a trailhead. A broad, level, gravel trail leads through the forest 0.4 mile (0.6 km) to an overlook before descending 100 feet (30 m) to the beach and stops 1 and 2.

17

A Spit in the Sea

DUNGENESS SPIT

Dungeness Spit protrudes into the Strait of Juan de Fuca like a skinny hitchhiker's thumb. The spit barely rises above sea level. At the highest tides, there is only 100 feet (30 m) or so of dry land separating the strait's surf and the shallow lagoon of Dungeness Harbor. The south coast of Vancouver Island forms the northern horizon, beyond the busy shipping lanes. In the evening the lights of Victoria sparkle across the water. On a clear day, ice-mantled Mount Baker juts skyward 75 miles (120 km) distant above the lighthouse at the tip of the spit. The 11-mile (18 km) round-trip hike to the lighthouse is a unique experience in the Pacific Northwest.

The spit is the product of longshore drift, the coastal migration of sand, gravel, and even larger stones eroded from unconsolidated Pleistocene glacial drift and interglacial deposits. The young, unconsolidated deposits form bluffs extending west along the coast for most of the 10 miles (16 km) to Port Angeles. Prevailing westerly winds in this region drive sea currents along the coast toward the east. Waves beating against the base of the bluffs undermine them, causing sediment to incrementally collapse onto the beach. The receding waves drag the sand and gravel straight down the beach slope. However, the incoming waves reach the beach at an angle due to the westerlies, so the next wave carries or pushes sediment

Most of the Dungeness Spit is protected as part of the Dungeness National Wildlife Refuge and is closed to public access; pick up a map at the trailhead kiosk, and please heed closure signs. The refuge and trail are open from dawn to dusk. There is a nominal fee per family at the trailhead. You may wish to consult a tide table and plan your trip to begin as the tide begins to ebb; avoid the highest tides. At high tides beach walkers will be doing a lot more log hopping on the spit, and waves may reach all the way to the bluffs to the southwest, which may not be safe to inspect. The beach is much wider at lower tides.

There is a public boat launch at Cline Spit. It is permissible to land a boat near the New Dungeness Lighthouse, but nowhere else within the refuge. A reservation in advance is required from the US Fish and Wildlife Service.

Dungeness Spit protrudes into the Strait of Juan de Fuca.

Looking north from Marine Drive, Dungeness Spit is a slender barrier separating Dungeness Harbor from the Strait of Juan de Fuca.

slightly farther east. The result is wholesale movement of large volumes of sediment downwind, or easterly, in a sawtooth pattern.

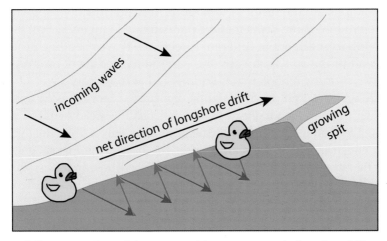

Sediment, mostly sand, is transported from west to east by longshore drift.

Spits form where shorelines change direction and the waves and wind transporting sediment lose energy and drop their sediment load. A submerged bar of sediment develops, extending offshore in the direction of the longshore currents, growing in length over time. Wind and particularly high tides coupled with storm surges, which usually occur in winter, deposit the above-water portion of a spit. Driftwood logs littered along the crest of Dungeness Spit attest to this high water. The logs trap blowing sand; if a fuzz of vegetation can take hold, more sand can be caught. As long as there is a sediment supply (and there is plenty in those western bluffs up-current of the spit) and the wind and sea conditions that started the process persist, spit growth continues. The limiting factors controlling the length of a spit are offshore water depth and stronger currents. Shoals off the northeast end of the spit extend underwater for another 0.5 mile (0.8 km) before the water depth exceeds 60 feet (18 m), but within another 300 yards (270 m) it reaches 600 feet (180 m).

Dungeness Spit is actually a complex of several spits, each heading in a different direction like a herd of cats. Graveyard Spit, named for the aftermath of an 1868 intertribal massacre, dangles from the southern side of Dungeness for more than 1 mile (1.6 km), reaching nearly back to the mainland. Only a narrow channel separates Graveyard from the little Cline Spit. There are other unnamed

slender spits; longshore drift along the lagoon shore of Graveyard
has built a slender appendage northward back toward Dungeness,
and another small, westward-growing spit encloses a small lagoon
on the southern side of Dungeness, east of Graveyard Spit. All these
minor spits are the result of variations in wind direction, spawning
different longshore drifts that are parasitizing Dungeness Spit itself,
moving its sediment in different directions. Waves bending around
the end of the spit's seaward end carry some sediment around the
corner, forming a hook that is submerged at most tides.

Dungeness Spit is a dynamic landform. Storms occasionally
breach the spit, generally at high tides, but the steady coastal process of longshore drift rebuilds the gaps after only a few tidal cycles.
Spit construction began shortly after the last of the Pleistocene ice
left the strait around 13,500 years ago. Recent research has shown
that the spit has grown about 15 feet (4.5 m) per year in length
over the past 120 years. At this rate, one might calculate that it
has taken less than 1,950 years for the spit to reach its full 5.5-mile
(8.8 km) length. But that does not account for the enormous volume of

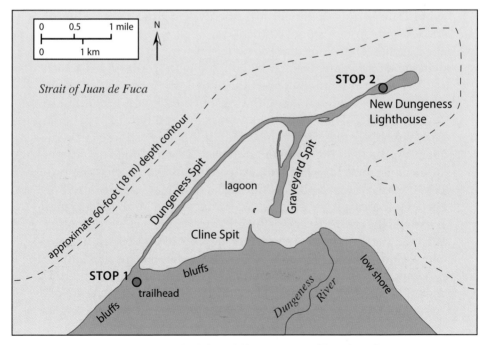

*Minor spits are built from drift processes working along the
shore of Dungeness or its dependent subspits.*

sediment, transported by longshore drift, submerged below the spit itself, nor does it account for the extent of the submerged platform of sediment, reaching northwest for nearly 1 mile (1.6 km). The spit has probably been forming throughout most of the Holocene, since the demise of the Puget Lobe of the Fraser glaciation (vignette 11).

Smaller appendages have come and gone, as well. For example, the slender, southwest-pointing minispit on the underside of Dungeness, just east of the junction with Graveyard Spit, has rebuilt itself in the last twenty-five years. The small spit formerly existed at that location but was not present at the time of a 1987 study. A 1990 aerial photo, however, shows the little spit just beginning to grow again; today, it is around 900 feet (275 m) long.

Local tourist brochures and some guidebooks call Dungeness Spit the longest spit in the world, though it is not even the longest in the United States; that title belongs to Washington's Long Beach Peninsula (see vignette 3), which is 22 miles (36 km) long. The longest spit in the world is probably the Arabat Spit, which parallels the Ukrainian shore of the Sea of Azov for about 66 miles (106 km). Dungeness Spit is nevertheless a classic spit. Its undeveloped, wild character and accessibility outweigh any hyperbole about its relative length.

Walk the forest trail to the shore. Once you have reached the sandy beach, it is tempting to set off along the seaward edge of the spit to the automated lighthouse at the end, 5.5 miles (8.8 km) away. Before you do, head southwest along the sandy beach, away from the spit, to the high bluffs (stop 1). The bluffs are composed of late Pleistocene deposits, the source of the spit's sediment. At first glance they look like homogeneous buff-colored sand, but the bluffs comprise some fine stratigraphy from the last glacial and interglacial periods.

The 100-foot-high (30 m) bluffs are mostly crossbedded gravel and sand. This is outwash, dropped by rivers flowing outward in advance of the Puget Lobe about 18,200 years ago. There are beautiful 3- to 12-inch-thick (8 to 30 cm) swirls and fingers of silt and fine sand reaching upward into beds of gray sand from a poorly sorted mix of silt, sand, and gravel. Called flame structures, they probably formed when the great weight of the overlying sediment and ice compacted the saturated, muddy deposit at the bottom and drove watery fingers of fine sediment upward into the gray sand. Faults offset thin beds in the sandy layer. Nearly at the bottom of the bluff are finer-grained, gray, clay- and silt-rich beds similarly deformed and mixed-up. If waves have scoured deeply enough, right at the base of the bluffs you'll see old compacted gravel, a fossil beach predating everything in the bluffs; the gravel looks rusty because iron

The bluffs at stop 1 contribute sediment to Dungeness Spit. The Olympic Mountains rise beyond. —Courtesy of Missy Rief

Flame structures extend upward from the brown layer into the overlying gray sandy beds.

oxide precipitated out of the overlying sediment. The iron oxide makes a good cementing agent between the clasts of this gravel.

The uppermost section of the bluffs is glaciomarine drift dropped from the floating portion of the Puget Lobe some 15,900 years ago. Enough water was taken up in glacial ice then that global sea levels were much lower, but the mass of all that ice also depressed the Earth's crust into the mantle slightly. There was a net sea level decrease of nearly 400 to 460 feet (120 to 140 m). The demise of the glaciers caused the world's oceans to rise, but the ice's removal also caused the crust to rebound upward. As an ice cube held beneath water in a bowl rises when you take your finger off of it, low-density crustal rocks float on higher-density mantle rocks (vignette 18). When thick ice sheets form on the crust, the solid mantle below slowly deforms and flows outward beneath the added mass, allowing the crust to subside.

After examining the layers in the bluffs, walk out along the spit as far as you wish. The lighthouse at the far end is stop 2, but you can just as easily appreciate the dynamics of the spit anywhere along its length. High tides will force you to walk on the soft sand above the waves or hop over logs. Remember to stay on the outer shore of the spit and out of the inner wildlife refuge, which officially extends east of the spit's crest; there are usually signs denoting the boundary.

If there is any kind of surf running, watch the wave patterns. Toss in a stick, or a rubber ducky, and track its movement. If conditions are typical, with the waves approaching the beach obliquely from the west or southwest, you'll see that stick get carried up the beach at an angle, then straight back down the beach. The next wave might carry the stick back up the beach, but on average it will end up farther east than where it started. You will need to be patient and watch for a while. This is longshore drift in action.

You may be surprised at the number of rounded cantaloupe-sized cobbles and boulders on the beach. It may be hard to picture, but currents and storm waves carried these clasts! They, too, were eroded from the bluffs and moved eastward bit by bit. They are found clear to the end of the spit.

Numerous shipwrecks and groundings have occurred on Dungeness Spit, which barely rises above the surface of the sea and is practically invisible in thick weather. The New Dungeness Lighthouse was completed in 1857 and has been automated since 1975. Volunteers staff it year-round. The station buildings are on foundations about 3 feet (1 m) aboveground, which is in turn about 6 feet (2 m) above the high tide levels.

Cobbles and boulders at the top of the beach attest to the power of waves. These have been carried about 2 miles (3 km) from the bluffs that end at stop 1.

Looking back from well out on the spit, you feel a long way from the mainland shore.

Is Dungeness Spit threatened by rising sea level? Will it someday be submerged? Consider the processes that created this landform. Despite probable submergence and destruction of the lighthouse buildings, longshore drift can be expected to continue. The bluffs that supply the sediment rise 100 feet (30 m) above the beach. As long as the bluffs are above the surf, they will continue to collapse and supply sediment to be transported eastward to the spit. It is reasonable to believe the spit will continue to exist, though if the rate of sea level rise is fast, storms will breach it more frequently, and completing the trek to the end could be problematic. So, my advice to you is don't wait. Go now.

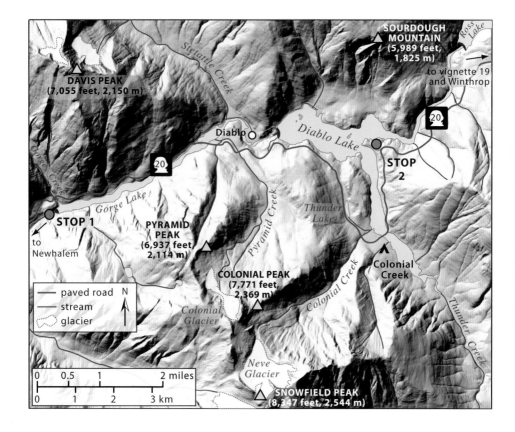

GETTING THERE: The Skagit Gorge is located between Newhalem and Diablo along Washington 20 in Whatcom County. The two stops of this vignette can be visited in either order, depending on the direction you are traveling. From the west, take Washington 20 from I-5 in Burlington. From east of the Cascades, travel Washington 20 west from Winthrop. Stop 1 is the well-signed Gorge Creek Overlook, about 3 miles (5 km) east of Newhalem. From the east, it is about 2.5 miles (4 km) west of the turnoff to the town of Diablo. Park on the west side of the bridge crossing the gorge. The short trail is wheelchair accessible. Stop 2 is the paved Diablo Lake Overlook, about 8 miles (13 km) east of stop 1, or about 0.2 mile (0.3 km) west of milepost 132. This trip could be combined with vignette 19.

18

The Deep Core of the North Cascades

Skagit Gorge

The Cascades north of Stevens Pass are characterized by steep and high rocky peaks, serrated ridges, deep glacial valleys, and cirque glaciers. To the south, all the way to Lassen Peak in northern California, the range is less dramatic. The few exceptions in the southern Washington Cascades are big solitary volcanoes, such as Mounts Rainier and Adams, and the granitic Stuart Range. The difference is largely due to the geology of the northern range, and the fact that the North Cascades have been repeatedly buried by thick sheets of erosive ice from Canada and episodes of local alpine glaciation far more extensive than today's. South of Stevens Pass, the range is principally Paleogene, Neogene, or Quaternary volcanic rocks. To the north, older metamorphic and plutonic rocks dominate, with sparse remnants of the volcanic rocks that once veneered the surface. Geologically the North Cascades are part of the Coast Plutonic Complex of the Coast Mountains of British Columbia rather than the largely volcanic Cascades to the south. The heart of the North Cascades is preserved in North Cascades National Park and the adjacent Glacier Peak Wilderness Area. Washington 20 (the North Cascades Highway) provides the most accessible route to the rugged interior of these mountains.

North Cascades geology is a jigsaw of rocks of widely divergent ages and sources. Their complexity hampered early geologists' efforts to understand the relationships between unrelated groups of rocks. Little systematic geology was done in the North Cascades until around 1950, and efforts continue into the twenty-first century. Through these mapping projects, essential rock units, faults, and structures were identified. New technologies, including the chemical analysis of microscopic mineral grains with an electron microprobe and isotopic dating, have advanced the understanding of these rocks and their complex relationships in time and space. The story of the evolution of the North Cascades and Coast Mountains of British Columbia is by no means resolved. There are several interpretations. All interpretations accept that masses of rock, or terranes, were added to the margin of North America by plate movements, but they differ in where and when this occurred and in the relationships between the several discrete groupings of rock making up this complex mountain range.

The North Cascades consist of around a dozen terranes. Each is composed of rocks related in age and depositional environment and that experienced a similar degree of metamorphism. Each terrane is separated from neighboring terranes by faults, and adjacent terranes may bear no relation in age, composition, or origin. The terranes originated at different times and in different parts of the Panthalassa Ocean, the world-girdling ocean that surrounded the Pangaea supercontinent. Today's Pacific is a remnant of this much larger ocean. Some terranes are pieces of ocean floor, others are alluvial fans deposited on a continental margin, and still others are the remnants of partially subducted volcanic islands.

The terranes were assembled by the extremely slow-motion convergence of tectonic plates and were attached, or accreted, to the

Looking east up Skagit Gorge. Newhalem lies at the foot of Davis Peak, left, and Pyramid Peak, right, both composed of Skagit Gneiss. At 9,066 feet (2,763 m), Jack Mountain, center top, is the highest point in the Methow Domain. —Courtesy of John Scurlock

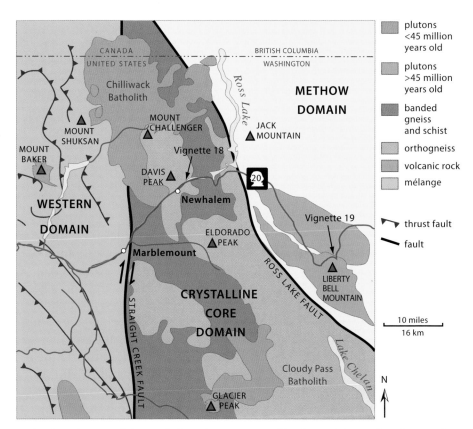

The three domains of the North Cascades are separated by the Straight Creek and Ross Lake Faults. Plutons older than 45 million years within the crystalline core were metamorphosed to orthogneiss. Younger plutons were not. (Adapted from Tabor and Haugerud 1999.)

margin of North America at different times and perhaps in different places. They converged with North America via the conveyor belt of the gigantic, oceanic Farallon Plate, now mostly subducted beneath North America (see vignette 16). The North Cascades terranes can be grouped into three different domains, separated by major north-south strike-slip faults. The three domains are distinguished from each other partly by the different degree of metamorphism that affected their constituent terranes.

The Methow Domain is the easternmost of the three. Until the mid-Cretaceous, long before the Cascades existed, the margin of North America was in the vicinity of the present-day Okanogan Highlands, in north-central Washington. Sediment washed into the Pacific Ocean to become the sedimentary component of the Methow Domain. It consists of folded but only slightly metamorphosed marine shale, chert, and sandstone and the basaltic seafloor rocks they were deposited on; ages range from 350 to 100 million years old. These rocks were compressed and deformed as North America drove westward during the breakup of Pangaea, accreting terranes along the way, including those in the heart of the Cascades and Wrangellia, a large terrane preserved in rocks of Vancouver Island (see the introduction). The Ross Lake Fault bounds the Methow Domain on the west.

The Western Domain forms the western margin of the North Cascades. The domain's eastern boundary is the inactive strike-slip Straight Creek Fault, a major structure that extends far northward into British Columbia. The Western Domain consists of a wild mix of clearly unrelated rocks called mélange. Individual rock units include subducted oceanic sedimentary and volcanic rocks of Jurassic and Cretaceous age; gneiss and igneous rocks of Paleozoic age; barely metamorphosed marine and volcanic sandstone and odd bits of limestone, all of Devonian and Permian age; and even ultramafic rocks from the mantle. The domain extends west to include the San Juan Islands. Some geologists refer to this domain as the San Juan Islands–Northwest Cascade Thrust System (after the thrust faults that stacked its terranes together and that today separate them) or the Northwest Cascades System. The degree of metamorphism, age, and origins of this domain's terranes varies considerably, indicating that the unrelated rock packages had experienced different geologic histories before they came together.

The Crystalline Core Domain, also known as the Metamorphic Core, is at the heart of the North Cascades, separated from the other domains by the Straight Creek Fault on the west and the Ross Lake Fault on the east. The crystalline core is the southern end of the gigantic Coast Plutonic Complex, which extends through the Coast Mountains and the Kluane Range (in southwestern Yukon). The complex is considered the world's largest belt of plutonic rocks. The accretion of Wrangellia with North America in the Late Jurassic or Early Cretaceous generated huge volumes of magma that intruded and metamorphosed the oceanic rocks that were caught in the middle of the collision. Much of the Crystalline Core Domain consists of the Skagit Gneiss. Gneiss is a metamorphic rock produced under great pressure and heat. Experiments on minerals found in Skagit Gneiss indicate that it was formed at pressures that correspond to burial 18 miles (30 km) deep in the crust and temperatures as high as 1,330°F (720°C).

There is considerable debate about where the terranes of the crystalline core converged against North America. Was it right where we find these rocks today, or did accretion happen elsewhere and the terranes then migrated northward along the Ross Lake Fault to their current positions? Adherents of the Baja BC hypothesis suggest that the Crystalline Core and Western Domains converged with North America as far south as the current latitude of Baja California, 900 miles (1,500 km) to the south of northwest Washington. This interpretation holds that the rocks then moved northward up the margin of North America by strike-slip faulting. Analogous tectonic movement occurs today along the San Andreas Fault, which is moving the western edge of California northward relative to the rest of North America. Other geologists are skeptical of the evidence and hold that convergence occurred where the rocks exist today, or with only modest northward movement. (See "E Pluribus Unum: Assembling Washington" in the introduction for more on this topic.)

In any case, the domains were assembled on the western edge of North America by the mid-Cretaceous, around 90 million years ago. Around 40 million years ago strike-slip faulting was replaced with renewed subduction off what would one day be the Washington coast. Magma invaded the crust, and the Cascade Volcanic Arc was initiated.

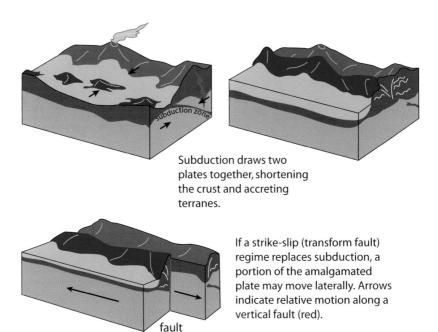

Subduction draws two plates together, shortening the crust and accreting terranes.

If a strike-slip (transform fault) regime replaces subduction, a portion of the amalgamated plate may move laterally. Arrows indicate relative motion along a vertical fault (red).

Accreted terranes can migrate laterally if subduction is replaced by a strike-slip, or transform, plate boundary.

Gneiss is defined by aligned, or foliated, mineral crystals that are coarse enough to be seen by the naked eye, and by the segregation of the minerals into light and dark layers or streaks. Crystals change their shape or orientation in response to pressure by elongating at right angles to the direction of pressure. Molecules in crystals migrate from pressurized zones to zones of less pressure within the same mineral crystal. The long axes of the resulting elongated crystals end up parallel to one another and impart a streaky appearance, or texture, to the rock if light and dark minerals are present.

The most familiar gneiss is banded, with alternating zones of dark- and light-colored minerals. These gneisses are derived from layered rocks. Much of the Skagit Gneiss lacks this highly visible layering because it was derived from plutonic rock. This type of gneiss is termed orthogneiss (from the Greek *orthos*, for "straight"). The elongated minerals may not be very apparent unless you examine the rock at just the right angle. The parent rock was tonalite, a salt-and-pepper intrusive rock similar to the more familiar diorite and granodiorite. The parental plutons intruded into middle levels of the crystalline core during a long interval in the Late Cretaceous to mid-Paleogene time, between 90 and 45 million years ago.

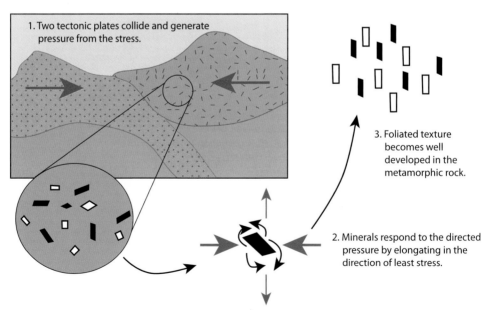

1. Two tectonic plates collide and generate pressure from the stress.

3. Foliated texture becomes well developed in the metamorphic rock.

2. Minerals respond to the directed pressure by elongating in the direction of least stress.

The foliated texture characteristic of metamorphic rocks results when minerals reshape and realign themselves in response to pressure.

Another ongoing debate among geologists revolves around the origin of the Skagit Gneiss. One model holds that the tonalite that would become the gneiss was buried deeply beneath stacks of accreted rocks during subduction. All these overlying rocks are now eroded. The other model says that huge volumes of magma intruded upward into the crust, which contained older tonalite plutons. The magma bodies flattened at their tops and ballooned laterally over the tonalite, pushing it deeper into the crust. Erosion has stripped away the plutons that cooled from the magma. In either case, the tonalite had to be buried as deep as 18 miles (30 km) in order to generate the heat and pressure necessary to metamorphose

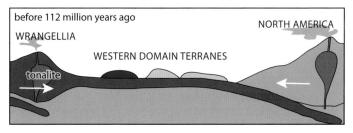

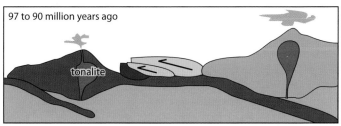

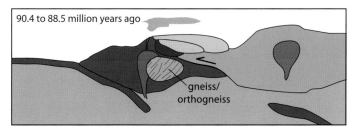

The collision model theorizes that compression during the collision of North America with eastward-moving Wrangellia thrust small intervening terranes westward. This buried the tonalite to the depths required to make high-pressure metamorphic gneiss and orthogneiss. (Modified from McGroder 1991.)

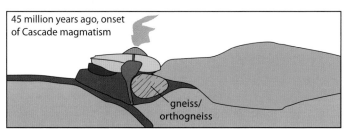

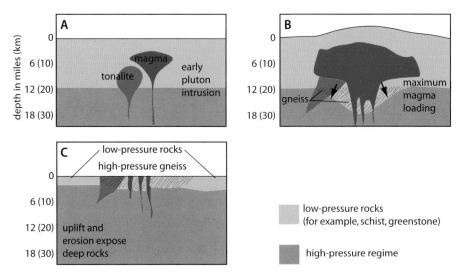

In the magma-loading model, (A) magma intrudes high in the crust and mushrooms above shallow host rocks (tan), including older tonalite plutons. (B) The added pressure pushes low-pressure rocks down into high-pressure, high-temperature conditions (gray), and these rocks become gneiss (pattern). (C) Some minerals exposed at the surface today formed at lower pressures, say that at 6 miles (10 km) below the surface, while in nearby rocks, often close to plutons, metamorphic minerals indicate that bands of the crust were carried much deeper to form gneiss. (Modified from Brown and Walker 1993.)

it into gneiss. There is adequate evidence to support the two theories, but not enough to disprove either. Both require much erosion of now-unknown overlying rocks.

The Cascade Range has been eroded only a short way into the North American Plate; the highest nonvolcanic peaks rise less than 2 miles (3 km) above sea level, and the deepest valleys are only 300 feet (90 m) or so above sea level. The total thickness of the plate beneath the mountains is around 25 miles (40 km). Yet rocks from deep within the crust are exposed on the surface. How? The answer is uplift, the balance between the rate of magma intrusion into the crust, erosion, and the relative densities of the continental crust and the mantle.

Hot, semimolten magma generated by subduction off the Washington coast rises upward into the solid continental crust because hot magma has a lower density—and therefore is more buoyant— than the crust it intrudes. A small portion of the magma erupts at volcanoes, but by far the greater volume cools within the crust as granitic plutons (vignette 19). These add to the volume and thickness of the continental crust. Because these rocks are lighter, the continental plates they compose float on denser rock in the mantle, just as a block of Douglas fir floats on water.

At the same time, erosion removes rocks at the surface and washes sediment out to sea, where much of it eventually leaves the continental plate and is carried far out onto the oceanic plate by turbidity currents (vignette 15). If erosion and intrusion rates are balanced, the crust maintains a constant thickness; however, the depth to any given bit of rock decreases due to erosion above. That bit of rock does not move upward through the crust relative to neighboring rocks. Rather, the contents of the crust as a whole migrate upward as the crust is replenished with magma below. If the rate of magma emplacement exceeds erosion, the crust thickens and rises ever higher above the mantle, and the mountains at the surface grow taller and taller. The result of this balancing act is uplift; it gives rise to mountain ranges such as the Cascades.

How fast does erosion strip away the surface? The Icy Peak pluton, east of Mount Shuksan, cooled an estimated 1.2 miles (2 km) below the surface 3.36 million years ago. These granitic rocks are now exposed at the summit of Icy Peak, elevation 7,073 feet (2,156 m). In that interval, the intervening rocks between the pluton and the surface eroded away. That is a rate of around 0.4 mile (0.6 km) per million years, or 0.02 inch (0.5 mm) per year. In the context of geologic time, this is very fast.

Subduction will continue to supply new rock to the core of the Cascades until the Juan de Fuca Plate has been entirely consumed beneath North America, some 20 million years from now. Then, erosion will outpace uplift, and, barring rearrangement of today's plates, the North Cascades will be ground down to low hills and finally to a broad plain—the geography that prevailed in pre-Cascades time when the Chuckanut Formation was deposited by rivers flowing all the way from the Rockies (vignettes 20 and 21).

At stop 1, the Gorge Creek Overlook, find the trail leaving the southeast corner of the parking lot, next to the toilets. The 0.2–mile-long (300 m) trail is paved (a little bumpy) and has a gentle grade. Turn a corner onto the slope above the river, with views up the Skagit Gorge; Gorge Dam is below. Turn right off the trail opposite a well-used gravel trail to the left. Walk through the trees 20 feet (6 m) to the outcrop. The rock is black-and-white-speckled Skagit Gneiss—very *nice* to look at (sorry—an old and very bad geology pun). At first glance you might think it was an unremarkable

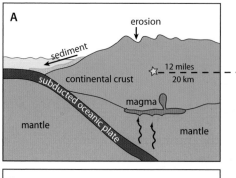

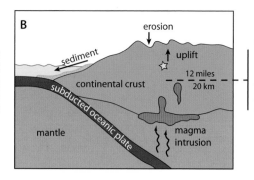

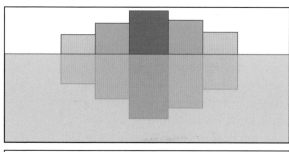

The star represents deep rock (A) at an arbitrary depth in the crust, say 12 miles (20 km). Rock at this depth migrates to the surface through combined uplift and erosion. Continental crust, consisting of relatively low-density rock, floats on denser rocks of the mantle. Magma rises out of the mantle due to subduction. Erosion strips the surface and sends sediment to the sea; turbidity currents carry it far out onto the subducted ocean plate. (B) Erosion has removed about 2 miles (3 km) of surface rock, and the rocks below have migrated higher as the continental plate is uplifted. The plate remains the same thickness, assuming magma intrusion is balanced with erosion. (C) With time, deep rocks reach the surface.

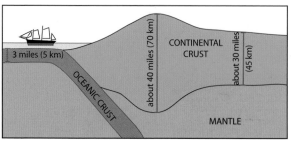

(A) Wood blocks of the same density but different thicknesses float at different heights in denser substances, such as water. (B) Crustal rocks behave similarly. Rocks in both oceanic and continental plates are less dense than mantle rocks, so they float on the mantle rocks. Ocean plates (basalt) are denser than continental plates (granite), so continental plates ride over oceanic plates in subduction zones. Thinner continental crust does not rise as high as thicker continental crust.

plutonic rock, a little coarser-grained than the Index granodiorite (vignette 14). Indeed, it was such a rock prior to metamorphism. But now the minerals are foliated.

The light minerals are quartz (glassy and smoky gray) and pla-gioclase feldspar (white). The dark ones are biotite, a member of the mica family. Your hand lens will help you see them more clearly. Find a blocky rock, not a thin slab, from the talus. Look for aligned minerals on one face and then rotate the block; the minerals on other faces do not appear elongated. Swirled white dikes of quartz and feldspar cut the gneiss in the outcrop. They were injected into the tonalite in the final stages of cooling, before metamorphism occurred. The dikes formed from the silica-rich fluid that was left in the tonalite magma after most of the other elements in the pluton had been incorporated in minerals that form at higher temperatures; quartz and feldspar are among the last minerals to crystallize in a pluton. The swirled appearance developed when the plastic, solid, but still-hot rock was metamorphosed, becoming gneiss. The gneiss was viscous like taffy as it formed, but it was not molten.

Back at the parking lot, walk onto the grated bridge over Gorge Creek and gawk at Gorge Canyon and the 240-foot (73 m) water-fall. During the Pleistocene glaciations, ice filled the ancestral Skagit

rotate 90°

No foliation is evident in one face (left) of this block of gneiss. White minerals are plagioclase and some quartz; the dark ones are biotite. When the block is rotated 90°, foliation is evident in the stretched and flattened minerals (horizontal in this photo). When the gneiss was deep in the crust, compression was perpendicular to the flattened minerals.

The swirled dikes at stop 1 invaded the tonalite as it cooled and were later deformed during its metamorphism to orthogneiss.

River valley. A tributary glacier occupied the gently sloping cirque high above on Mount Ross. The Skagit glacier eroded a U-shaped valley into the gneiss. Once the glacier melted away, the Skagit rapidly cut downward, making the modern V-shaped Skagit Gorge. Gorge Creek, flowing out of the cirque, encountered shattered rock in a fault; it was more easily eroded than the solid gneiss. Gorge Creek became trapped in the broken rock and followed the fault's trace down the steep valley wall, racing to keep up with the ever-deepening Skagit Gorge. The ancient fault can be traced on the opposite side of Skagit Gorge, but it is less pronounced and not eroded into a canyon. The fault is not an active structure, so it does not present a hazard to Gorge Dam.

Now proceed to stop 2, the Diablo Lake Overlook. Though you are in the middle of a spectacular mountain range, the elevation here is only 1,700 feet (518 m). The peaks in all directions except the east are eroded into Skagit Gneiss. The main attraction, other than the gorgeous scenery, is the spectacular vertical roadcut across the highway. Cross carefully and push through the little trees for a close look. Many white dikes cut the gray orthogneiss. This rock is migmatite, a mix of gneiss and igneous dikes. The Skagit Gneiss here is

At stop 2, Colonial Peak, left, and Pyramid Peak, center, soar above the green water of Diablo Lake, colored by glacial melt. The peaks are composed of Skagit Gneiss.

White dikes intrude dark orthogneiss in the 50-foot-high (15 m) roadcut at stop 2. Note the man for scale, lower left.

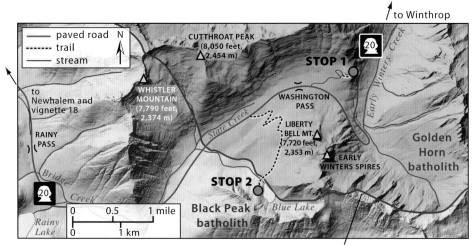

contact between Golden Horn and Black Peak batholiths

GETTING THERE: Washington Pass is the highest point on Washington 20 (North Cascades Highway). From the west side of the state, the pass is about 100 miles (160 km) east of Burlington off I-5. From the east, the pass is about 30 miles (48 km) west of Winthrop. The elevation of the pass is 5,477 feet (1,669 m).

Stop 1 is the viewpoint just north of the pass. Stop 2 is at Blue Lake, reached by a moderate 4.4-mile (7 km) round-trip hike on the popular Blue Lake Trail. The well-signed trailhead is about 1 mile (1.6 km) west of the pass. A Northwest Forest Pass is required for parking. Blue Lake lies at 6,250 feet (1,905 m), and the elevation gain to the lake is 1,050 feet (320 m).

Three different plutonic rocks. Dark diorite, rich in mafic minerals, and pinkish-yellow, silica-rich (felsic) Golden Horn granite rest on a typical granodiorite or tonalite boulder. Quartz fills a fracture in the granodiorite.

The sunlit east face of Liberty Bell Mountain, right, rises a sheer 1,000 feet (305 m) just south of Washington Pass. Peaks to the left are the Early Winter Spires.

The crystals in this freshly broken sample of Golden Horn granite are around 0.08 inch (2 mm) across. The knife tip is 0.4 inch (1 cm) long. Pinkish-yellow orthoclase dominates, and a smooth crystal face glints at bottom right. Quartz crystals are smoky or clear. The black mafic mineral is arfvedsonite, an amphibole mineral similar to the more familiar hornblende; the one at lower right shows the diamond-shaped cross section typical of amphiboles.

A pluton results from the cooling of an individual body of magma within the crust. A batholith develops when repeated magma intrusions place numerous plutons together in the crust. Younger pulses of magma typically overlap or intrude earlier ones. Batholiths can be enormous. The Golden Horn batholith is a small one, roughly 175 square miles (450 km²). The Chilliwack batholith, north and west of the Skagit Gneiss (vignette 18), is 380 square miles (980 km²). Both are among the many batholiths in the gigantic Coast Plutonic Complex, which extends up the British Columbia coast into Alaska.

The overlook at stop 1 is perched on the brink of a granite cliff, 700 feet (215 m) above Washington 20. The rock you are standing on, which forms all the peaks in view, is composed of Golden Horn

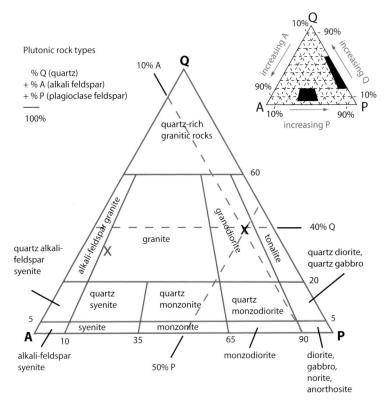

Plutonic rock types

% Q (quartz)
+ % A (alkali feldspar)
+ % P (plagioclase feldspar)
————
100%

Plutonic rocks are classified by the relative percentages by volume of quartz (Q), potassium- or sodium-rich alkali feldspar (A), and plagioclase (P) minerals, totaling 100 percent. Other minerals are not counted. The triangle at upper right shows the grid pattern; fields of tonalite and quartz monzonite are blacked out as examples. In the lower triangle, red lines define compositional boundaries for named rocks; the lines need not parallel edges of the triangle. Numbers refer to mineral percentages, as shown in the upper grid. Identify and count these minerals in a rock, plot the percentages, and you'll learn its name. A rock with 40 percent quartz, 10 percent alkali feldspar, and 50 percent plagioclase (blue lines) plots at the black X and is granodiorite. Golden Horn samples plot at the green X.

granite. The Golden Horn itself is a peak about 6 miles (10 km) northwest of stop 1. When Golden Horn granite is weathered, the orthoclase assumes a golden color. Liberty Bell Mountain (7,720 feet, 2,353 m), along with its neighboring spires directly south of the overlook, is a rock-climbing mecca. If the weather is decent and you have binoculars, you will likely see climbers scaling the clean, solid, glacially smoothed granite. Across the valley of Early Winters Creek, to the east, are more peaks made of Golden Horn granite.

At 8,876 feet (2,705 m), double-peaked Silver Star Mountain is the highest peak in the region. To the left are the four matched towers of the Wine Spires. All consist of Golden Horn granite. Erosion along joints and faults cuts deep notches and gullies to isolate the spires. —Courtesy of ©Steph Abegg

Climbers ascend Liberty Crack on the east face of Liberty Bell Mountain. The Golden Horn granite provides some of the most classic rock routes in the Cascades. —Courtesy of ©Steph Abegg

Glaciers polished the rock slab at the overlook. Grit encased in the ice made it smooth and shiny as the ice slid past, not the soft ice itself. Small stones dragged along by the ice left striations. Weathering has roughened the surface, removing much of the polish.

Glacial polish reflects in the sun at stop 1. Glacial striations run diagonally in front of the man's toes.

All this granite is a huge mass of rock that cooled from magma. Based on mineral size and chemistry, it cooled about 2 miles (3 km) below the surface. Uplift (vignette 18) brought it to the surface to face the ravages of erosion and weathering. Where did such a huge volume of magma come from? Did it begin as a single, large volume, or did it accumulate in the crust over time? Geologists continue to puzzle over these questions.

Geologists theorize that magma in a subduction zone, such as that beneath the Cascades, originates in the mantle wedge, the part of the asthenosphere that lies between the subducted oceanic plate and the crust. The addition of water, carried into the mantle in the saturated ocean floor basalt and blanket of sediment, triggers the mantle to melt; the melting point of solids is lowered by the addition of water. The resulting magma, being both hotter and more liquid than the surrounding rock of the asthenosphere, is less dense, so it rises.

But in what form the magma rises is a bugaboo for researchers. The classic model says that magma moves upward as a diapir, or blob of molten rock, like the hot wax in a lava lamp. The volume of magma moving into the crust in this model presents a problem: there aren't huge voids the size of plutons in the crust, so something has to make way for the magma to migrate upward. This "room problem" has driven theories for magma emplacement over the past century, and the debate continues.

So how could a diapir make its way upward? At 6 to 20 miles down (10 to 30 km) in the crust, where temperatures exceed 570°F (300°C), rocks are hotter and less viscous. A large batch of magma could push upward slowly as ductile rocks flow downward along its sides and fill in behind the trailing tail of the diapir; however,

The hot red wax in a lava lamp rises because the wax is heated by the bulb at the bottom to become slightly less dense than the clear oil. When the wax reaches the top of the lamp, it cools, grows denser, and sinks. One difference with rising magma is that rock doesn't sink as it cools.

rising magma of this volume must stall when it reaches more-brittle rocks closer to the surface. So how is it we find plutons surrounded by brittle country rock — the rock that exists closer to the surface?

The diapir may balloon out along the brittle-ductile boundary in the crust. Smaller volumes of less-viscous magma may continue upward as dikes following fractures, some of which may be caused by the pressure of the diapir. Blocks of overlying country rock, fractured by the rising magma, may fall into the diapir, allowing portions of the diapir to continue rising. Being cold and dense, the country rock could conceivably fall all the way through the magma, or melt and be assimilated.

Part of the problem is that no one has ever observed a rising diapir, or any other migrating magma body. We know that magma can accumulate in huge volumes and cool as plutons. There is good field evidence to support the diapir model. The country rock surrounding some plutons is deformed, its minerals are foliated near the contact,

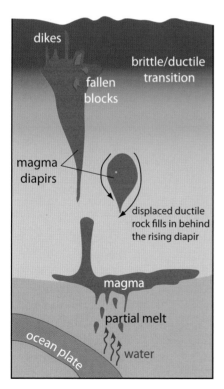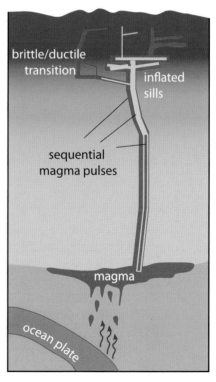

Two models for the emplacement of large volumes of magma in the crust. Diapirs (left) hypothetically stall and balloon outward at the brittle-ductile transition in the crust. Dikes (right) stall and inflate with new magma pulses.

and the deformation decreases with distance from the contact. This would indicate the large blob of magma was emplaced in one event. Blocks of surrounding country rock, sometimes partly melted, are commonly found in plutons. The variation in the chemical composition of adjoining plutons may be due, in part, to the incorporation of rocks they intrude.

That said, the model of a single massive diapir may be overly simplistic. A growing number of geologists have begun to accept a model suggesting plutons grow incrementally. It posits that magma moves upward in much smaller batches via long dikes and accumulates piecemeal in interconnected dikes and sills. These structures inflate as sequential batches of magma arrive. This model minimizes the room problem by reducing the size of the rising magma body. The evidence for this theory is found in the ages of many plutons, which show that they grew over several million years. This lengthy formation would only be possible if small magma batches were repeatedly injected into a pluton, keeping the magma close to a molten state over such a long time. Most likely, a combination of these processes occurs.

On the way to stop 2, look east at Liberty Bell Mountain and its neighboring spires. The deep notches between the peaks are parallel faults trending east-northeast. Movement occurred along the faults when the pluton was shallow in the crust and brittle, leaving shattered rock on either side. As the pluton reached the surface via regional uplift, the broken rock eroded, resulting in the notches. Along the trail to Blue Lake you'll notice that the granite is crumbly. Sharp-edged crystals have weathered out of outcrops and accumulated in heaps called grus. Repeated freezing and thawing expands and contracts the water that has penetrated the granite, wedging its crystals apart. The loose crystals provide a great opportunity to use a hand lens.

Blue Lake is set in a rock-walled cirque eroded by glaciers, though no ice remains today. Walk the trail across the lake's outlet and gain the knob above the northwest shore. A high ridge encircles Blue Lake on all sides but the north. To the east Golden Horn granite extends to the 7,800-foot (2,380 m) unnamed peak directly above the lake, but the high wall above the big talus slope south and west of the lake is 91-million-year-old (Cretaceous) granodiorite of the Black Peak batholith. This mass of plutonic rock intruded the suture zone between the accreting Insular Superterrane and the crushed seafloor rocks that were caught between it and North America (see "E Pluribus Unum: Assembling Washington" in the introduction). The Golden Horn magmas intruded the eastern margin of the Black

Weathered granite and grus along the Blue Lake Trail.

Looking east from Blue Lake. Deep notches between (left to right) Liberty Bell Mountain, Concord and Lexington Towers, and the Early Winter Spires are eroded faults. The dark rock is weathered and veneered with algae and lichens.

Peak batholith about 40 million years later. The contact between the two batholiths runs through the saddle east of the lake. You should be able to find two colors of talus at the bottom of the eastern slope, near the lake; hike cross-country along either shore to get there.

The Black Peak talus is salt-and-pepper tonalite, and very different in appearance from the Golden Horn granite. There is a higher proportion of plagioclase and very little, if any, orthoclase. Mafic minerals are the by-now familiar hornblende and biotite. The rock may have a greenish tinge, especially along fractures. These minerals are epidote and chlorite, resulting from the chemical alteration of mafic minerals and plagioclase when hot water permeated the rock, before it was uplifted to the surface.

Your encounter with the Golden Horn granite may not have shed much light on the problem of pluton ascent and emplacement. It takes observations at many plutons to begin to make any generalizations about those vexing problems. But your time at Washington Pass does provide an opportunity to study this unusual Cascades granite up close and from afar, and to marvel at the landforms produced when crustal magmas are exposed to the ravages of erosion.

Gray granodiorite walls of the Black Peak batholith above the south shore of Blue Lake are notably different in color from granite of the Golden Horn batholith. — Courtesy of Keith Wyman

20

Honeycomb Weathering
LARRABEE STATE PARK

Along the rocky shore of Larrabee State Park, sandstone of the Eocene-age Chuckanut Formation weathers into an intricate fretwork of small cavities separated by thin walls due to a bizarre concurrence of geologic and botanical processes. The cavities are up to 6 inches (15 cm) deep and are often aligned in rows or develop in dense concentrations. The combination of deep shadows and yellow light on the tan sandstone is particularly beautiful in low, slanting afternoon light.

The honeycomb structure is known as tafoni. The origin of *tafoni* (singular *tafone*) is not clear. It may be derived from the Greek *taphos*, for "tomb," but I'll put my money on an alternate origin, from either of two Corsican words: *taffoni*, meaning "windows," or *tafonare*, meaning "to perforate." *Tafoni* is also Sicilian for "windows." The first scientific descriptions of tafoni were made in the mid-1800s based on outcrops in central Spain, where tafoni occur in granitic rocks, and later in Corsica.

Let's first differentiate between weathering and erosion. Weathering is the loosening of minerals or clasts that compose a rock. It can be a physical or chemical process. Physical weathering occurs because of pressure. There are quite a number of physical weathering processes. During alternating freeze-thaw cycles, water trapped in rock expands and contracts, loosening crystals and clasts. Blowing sand or waterborne sediment loosens grains in rock, as does microscale fracturing caused by the tiniest of roots growing into a rock surface.

Chemical weathering involves changes to the molecular structure of the minerals that compose a rock. Water added to a crystal's molecular structure weakens the bonds holding atoms together, so it is easier for ions and molecules to be dissolved in water. Rain may contain carbon dioxide from fossil fuel combustion, forming weak carbonic acid, or acid rain, which weakens the bonds in calcium-rich rocks, such as limestone and marble. This softening of the rock allows physical weathering to work at a greater rate.

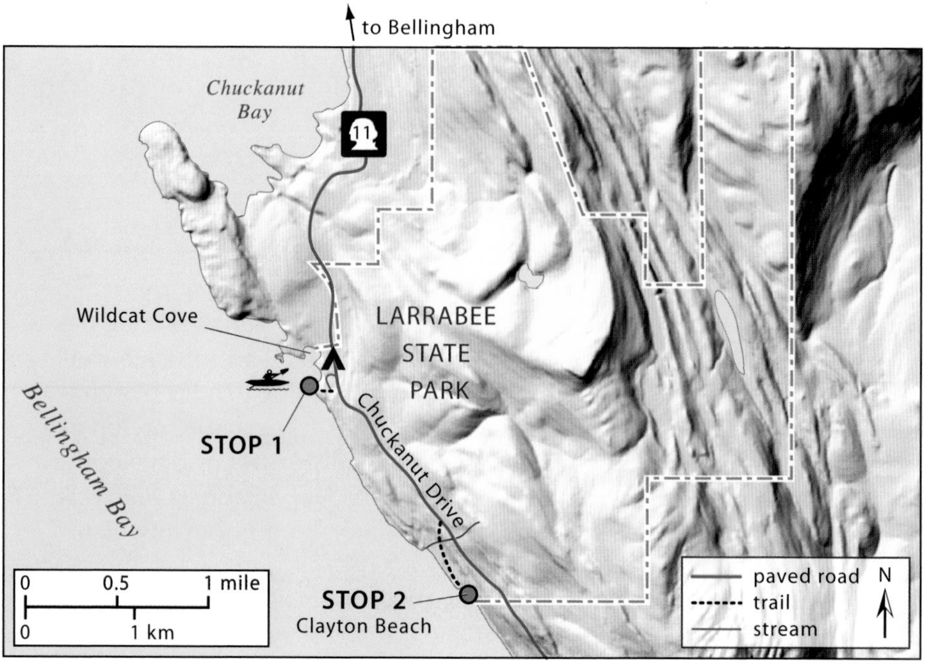

to Bellingham

Chuckanut
Bay

11.

Wildcat Cove

LARRABEE
STATE
PARK

Bellingham Bay

STOP 1

Chuckanut Drive

| 0 | 0.5 | 1 mile |
| 0 | 1 km | |

STOP 2
Clayton Beach

paved road   N
trail
stream

GETTING THERE: At Bellingham, take exit 250 from I-5 and go west on Old Fairhaven Parkway (Washington 11) to the light at 12th Street. Turn left and follow scenic Chuckanut Drive (Washington 11) 5 miles (8 km) to the entrance to Larrabee State Park, on the right. Go past the entrance kiosk and turn left, following signs to the picnic area and the parking above the band shell. Walk to the beach, passing under the railroad tracks in a tunnel painted with scenes of local wildlife. At the foot of the steps, turn right and walk down the trail 250 feet (75 m) to the beach. The outcrop on the right is stop 1.

To reach the Clayton Beach Trailhead and stop 2, return to Chuckanut Drive and proceed another 0.5 mile (0.8 km) south to the signed parking lot on the left. Cross the road and walk the easy but frequently wet trail 0.5 mile (0.8 km) to the railroad tracks. The trail descends a very short section of rock slab just before reaching the railroad tracks. Angle left across the tracks and follow the trail that is more or less parallel to the tracks to reach the big sandy beach.

You can examine the intricate rock structures discussed in this vignette at either stop. If you prefer a longer walk, go directly to stop 2, which has the bonus of a sandy beach—an infrequent occurrence on the Salish Sea's coast. A Discover Pass is required to park at either of these stops.

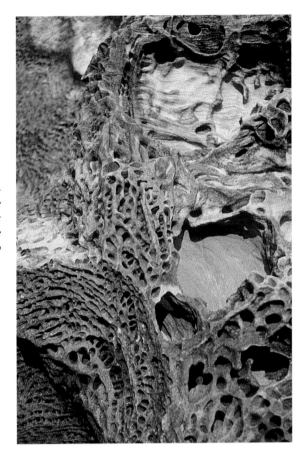

*Honeycomb weath-
ering in sandstone
produces complex
rock surfaces. The
cavities are also
known as tafoni.*

Erosion, on the other hand, is the physical removal of weathered rock. The same water (streams, ocean waves, or even hard rain) or wind that loosens surface grains will carry them off. Gravity causes rock fragments to move downslope. Glaciers do their bit, too (see vignette 12).

Stop 1 is the rocky point at the north end of the little beach. The honeycombs are near the base of the rock wall on a wide ledge, just above the high tide line. Use caution here if the rock is wet. There are more examples around the corner.

Honeycomb weathering requires three fundamental conditions: permeable rock, a source of soluble salt, and repeated episodes of wetting and drying. In coastal environments, the complex honey-comb structure begins with salt water splashing onto the rock above the high tide line. As the rock dries out, salt crystals grow between the grains composing the rock. The sandstone of the Chuckanut

The honeycombs at stop 1 are on the ledge just to the left of the man.

Formation is an ideal host for this process because the pore spaces between grains hold water.

The more slowly the salt water dries, the larger the salt crystals can grow. The crystal growth is the form of physical weathering that loosens the grains. The salt crystals may be dissolved the next time a wave splashes the rock, but the process then begins again, slowly prying the rock apart. Small pits in the rock are preferentially enlarged because they hold more water and dry more slowly, therefore allowing salt crystals to grow larger. There is also a chemical component to salt weathering: the crystalline structure of some minerals, such as biotite and plagioclase feldspar (very common in Chuckanut Formation rocks), are more susceptible to saltwater absorption than others, such as quartz, so they are more easily broken apart and wedged from the rock.

In temperate coastal zones, such as western Washington's, endolithic algae readily colonize rocks, shells, coral, and even the spaces between individual grains of rock. You may have an unintended encounter with these plants if you slip on wet rock along the shore, so watch your step. Research at the microscopic level has uncovered fascinating things about the relationship between tafoni and algae. The dark coloring of the walls dividing some cavities is a thin coating of microscopic algae, contrasting with the "cleaner" sandstone at the back of the cavities. Salt crystals coat the algae, but no salt has been found beneath it, indicating the algae forms a protective film on the rock, inhibiting saltwater intrusion. You can see for yourself the effect algae has on water absorption. Just drip some water on a well-defined dark rock surface and time the absorption. Do the same on bare rock. Be scientific about it, and repeat the experiment several times. One study found that bare Chuckanut cavities absorb water in one second, whereas the darkened sidewall rock took fourteen seconds.

When salt weathering and algal growth occur at a balanced rate, tafoni beginning as small pits can grow into the rock, and the expanding film of algae protects the rock between cavities from further salt weathering. The algae can grow along the sidewalls and into the cavities, and salt weathering is concentrated at the back

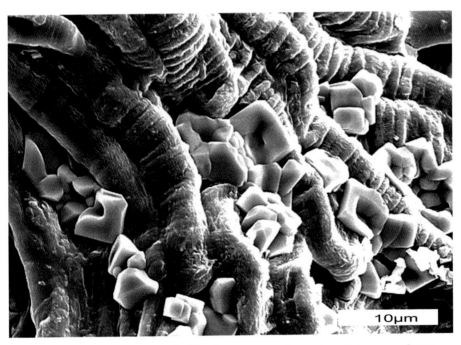

An electron microscope image of cubic salt crystals growing on algae living on sandstone. The algae reduces the amount of salt penetrating the rock, thus inhibiting salt weathering. The scale bar is 10 microns (0.00004 inch, 0.001 mm). —Courtesy of George Mustoe

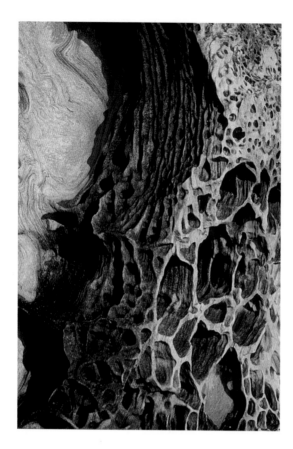

A dark film of algae covers some of the rock at Clayton Beach (stop 2). The ridges of the honeycomb weathering parallel the sandstone's bedding, top center, which has been tilted to vertical. The bedding is also visible in some tafoni, right.

of the cavity where algae have not spread. If weathering and erosion exceed colonization rates, cavities expand more rapidly laterally because the sidewalls aren't protected. Continued weathering and erosion may cause the incipient cavities to coalesce with their neighbors, making a relatively smooth rock surface. At the other extreme, if algal colonization exceeds salt weathering, weathering and erosion are retarded and the surface of the rock is stabilized. All of these scenarios can be seen on a single rock face at Larrabee State Park. There are surfaces completely covered with dark algae, others that are profusely honeycombed, and still others that are algae-free and fairly smooth. You may even see tiny white salt crystals growing on a rock surface that you can easily brush off with your fingers.

Stop 2 offers a wider variety of honeycomb features. Once you reach the big sandy beach, walk to the north. Especially fine tafoni can be found on large sandstone blocks isolated from the main cliff

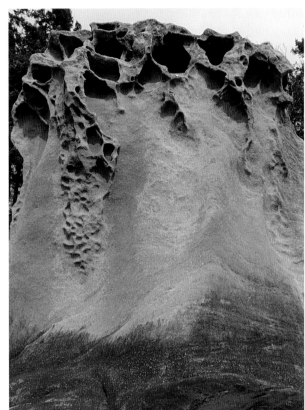

The pace of salt weathering has exceeded algal growth in the center of this 10-foot-high (3 m) Clayton Beach rock face. The large, intact tafoni at the top give an idea of how much rock has eroded from the smooth surface below. Green algae completely cover the bottom of the rock face.

at the north end of the big beach; walk around these to the seaward side. In places the tafoni are aligned with the nearly vertical cross-beds. This might be because mineral grains deposited in these particular layers are more prone to absorbing salt water. Chuckanut Formation sandstone is arkosic, meaning there is a high proportion of plagioclase feldspar along with the quartz that is ubiquitous in sandstone. Arkosic minerals are more susceptible to salt weathering than quartz.

The Chuckanut Formation was deposited by rivers flowing across a floodplain in the Eocene epoch; the rocks along the beach are about 57 million years old (see also vignette 21). The sandstone consists of layers of sand-sized rock clasts and minerals. During periods of more energetic flow, the rivers occasionally scoured out sand previously deposited on their beds, truncating horizontal layers and depositing layers of sand on top of and next to the truncated ones. With each new surge of current, older layers were truncated and new layers were deposited at different angles. This sort of layering is called crossbedding. The Chuckanut Formation was folded around 40 million years ago, likely due to the uplift of the Crescent Formation basalt and the accretion of the Olympic Subduction Complex to the margin of North America (vignettes 15 and 16). The folding is evident in the steeply dipping crossbedding in the crags along the beach.

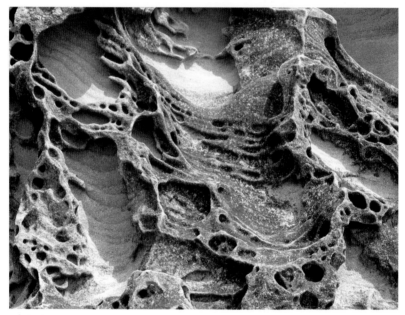

The honeycomb structure runs parallel to sandstone crossbeds. Where the dark algae are not present, salt weathering has outstripped honeycomb development.

Tafoni are well developed in Chuckanut sandstone in the San Juan Islands, and in similar sandstone of the Nanaimo Formation in the Gulf Islands, just across the international boundary. They are found in sandstone on the Olympic coast and in the Crescent Formation basalt along Hood Canal.

Though the lacy honeycomb structure is found in other rock types, including basalt, granite, and rhyolite, it is not found in shale and metamorphic rocks, which are relatively impermeable. Honeycombs also develop in noncoastal environments. Algae are not necessary for their formation, though honeycombs that develop without algal protection may not be as stable. Honeycomb weathering occurs in desert environments, both cold and hot. In deserts

Honeycomb weathering can develop in any permeable rock, including this Crescent Formation basalt at Tongue Point, west of Port Angeles. —Courtesy of Scott Babcock

far from the sea, the salt may leach out of the rock itself or be dissolved in water that permeates the rock and leaves salt to begin the weathering process. Tafoni have been found from the sun-parched sandstone canyons of the American Southwest to the frigid Dry Valleys of Antarctica.

Honeycomb weathering can be a significant contributor to the slow, steady erosion of coastal crags. A study conducted on coastal sandstone in California determined that cavities can, in a geologic sense, deepen rapidly, as much as 0.06 inch (1.4 mm) per year; however, 0.004 to 0.02 inch (0.1 to 0.6 mm) per year was more typical. At the slower rate, between seventy-five and four hundred years must pass for an individual pit to become 1.6 inches (4 cm) deep. A quarry face cut in Chuckanut sandstone in 1903 had developed 2-inch-deep (0.65 mm) honeycomb structures by 1981, meaning they had grown at about 0.03 inch (0.65 mm) per year. In the Arizona desert, the rate has been shown to be far slower: 0.002 to 0.0005 millimeter per year in sandstone and only 0.0002 to 0.00005 millimeter per year in dolomite, a calcium-rich rock similar to limestone.

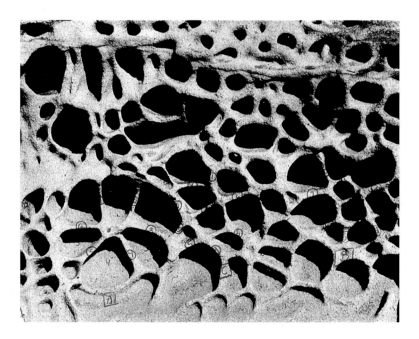

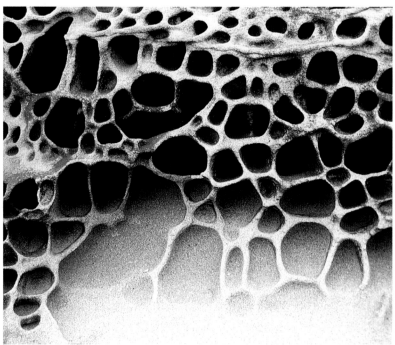

These photos were taken at different times (1975, top, 2005, bottom) to study weathering rates in coastal sandstone at Pebble Beach, California. Notice how some cavities have disappeared and many of the walls dividing cavities grew much thinner. —Courtesy of ©Jon Boxerman

Some rock climbers use tafoni as handholds along this shoreline. The thin walls between the honeycombs, however, are easily broken, and climbing on them should be discouraged. A single broken foothold can erase hundreds of years of development. Honeycomb weathering along the Chuckanut shore is a natural feature cherished by many visitors. The beautiful shapes and shadows in the golden light of evening are the inspiration for many paintings and photographs.

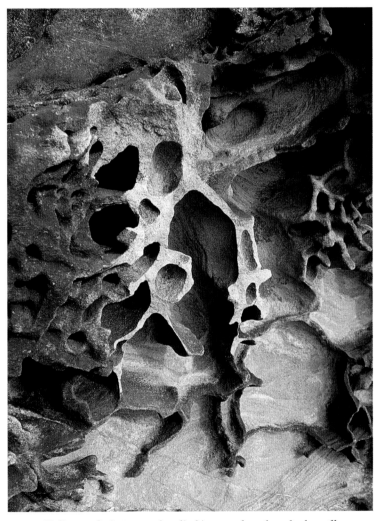

Delicate tafoni at a popular climbing area have been broken off.

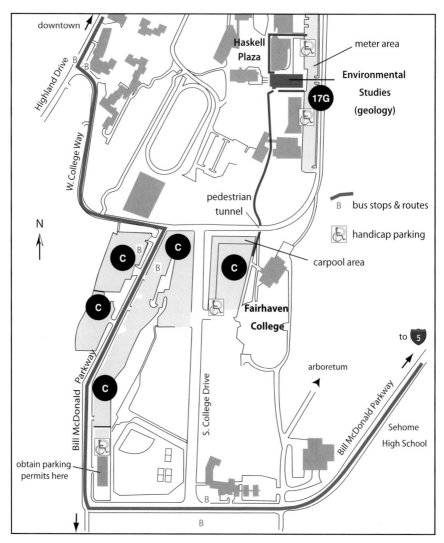

Map of the south end of the Western Washington University campus showing routes (blue lines) to the Environmental Studies building, each of which is wheelchair accessible.

GETTING THERE: The Western Washington University Geology Department's displays are in the corridors of the Environmental Studies building. From I-5 in Bellingham take exit 252 and follow signs to the university via Bill McDonald Parkway. Whatcom Transportation Authority buses visit the campus with regularity. If you must drive, the gravel C lots on either side of South College Drive, at the south end of campus, do not require permits after 4:30 p.m. Monday through Friday or all day on weekends. At other times, purchase a parking permit at the parking office at the southwest corner of campus (21st and Bill McDonald Parkway). The Environmental Studies building is 0.2 mile (0.3 km) north of the C parking lots. Expensive metered parking is available at all times in lot 17G, immediately east of the building.

21

The Big Bird of Western Washington

DIATRYMA

Geologic displays line the corridors on three floors of the Western Washington University Geology Department. They are free and open to the public. One highlight is an exhibit of fossil footprints preserved in the Chuckanut Formation, including spectacular tridactyl (three-toed) tracks made by the giant flightless bird *Diatryma*. Tracks of this bird, twice the size of ostrich tracks, can be seen nowhere else in the world. In its day *Diatryma* was one of the largest terrestrial animals on the planet.

Around 52 million years ago, during the Eocene epoch, much of the area that now lies east of the Salish Sea was a coastal floodplain. Rivers meandered back and forth across the landscape, occasionally flooding and depositing fine mud. These rivers began on the slopes of the Rocky Mountains; the sediment they carried across what would later be Washington were eroded from those mountains, recycled, and redeposited in subsiding basins near the coast. The shore of the Pacific Ocean was about where the Olympic Peninsula rises today; the rocks of the Olympic Mountains were still below the sea (vignette 16). The sediments in the northern part of the floodplain would become the rocks of the Chuckanut Formation, today exposed principally in the foothills of the Cascades in Whatcom County, west of Mount Baker, and along the shore at Larrabee State Park (vignette 20).

Geology displays are accessible seven days a week, including evenings (except for intersession breaks and university holidays). As of this writing, all doors are open on weekends. There are elevators in the building. Due to a flight of steps along the west side of the building, visitors in wheelchairs coming from parking lots C or 17G will need to enter the ground floor on the building's south side; alternately, you can use the handicap parking area at the north end of lot 17G (metered or with permit) and reach the building from the north through Haskell Plaza. This trip makes a great outing on a dreary winter day. Bring the kids!

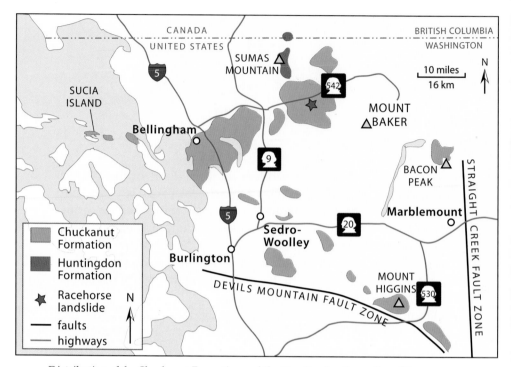

Distribution of the Chuckanut Formation and the Huntingdon Formation of the same age.

Subtropical forests formed a nearly unbroken carpet along the floodplain. Temperatures were warm here. Climate reconstructions, using leaf fossils found in the Chuckanut Formation as evidence, indicate a mean annual temperature of 71°F (21.7°C) in what would eventually be known as western Washington, equivalent to today's Okefenokee Swamp or Kona, Hawaii. In contrast, the mean annual temperature at low elevations in coastal Washington today is 51°F (10.6°C).

The dinosaurs had died out a mere 13 million years earlier. Mammals and birds were now the dominant land animals. Except for a single turtle shell, no fossilized bones have been found in the Chuckanut Formation; however, a number of animal species left tracks in the mud. Leaf fossils are abundant in the Chuckanut. Two of the most common are large fronds of palms of the *Sabalites* genus, and *Cyathea pinnata*, a tree fern. Some Chuckanut shale beds are as much leaf as rock. Comprising essentially pure organic deposits, these beds are indicative of plant-rich swamps. Coal is common in the Chuckanut Formation.

The Chuckanut is rich in exquisite Eocene-age plant fossils. Top to bottom: the tree fern Cyathea *(with a modern sword fern), wood grain preserved in a petrified log, and* Glyptostrobus *(water pine).*

George Mustoe, the paleontologist responsible for the Geology Department displays, stands next to latex casts of Sabalites *palm fronds of the Chuckanut Formation. The modern descendant of this tree is the common sabal palm, or palmetto, a native of the southeastern United States, Caribbean, and tropical South America.*

Picture the scene one can interpret from the Chuckanut Formation's composition and fossils, including its trace fossils, or tracks. It is another hot, muggy day on the floodplain 53 million years ago. It's no different from yesterday, or last month, or several months ago. And it won't be different tomorrow, or next month, or the next 1 million years. Palms and ferns and other semitropical plants crowd the shores of a sluggish river meandering in great loops. Needle-billed herons stalk prey. A turtle basks on a floating log. Herds of dainty *Hyracotherium*, standing all of 10 to 18 inches (25 to 45 cm) at the shoulder and 2 feet (0.6 m) long, munch on the tough leaves of low bushes. They are among the earliest relatives of the horse. Toothy mammalian predators called creodonts watchfully eye the little prehorses. Fat hippo-like animals sink deeply into the mud, leaving behind many deep but nearly shapeless sets of tracks.

A set of fresh, bipedal three-toed tracks slowly fills with water on the shore; the heavy animal that made them sank 1 inch (2.5 cm) or more through the thin veneer of mud and into the firm sand beneath. The 10-inch-long (25 cm) prints, nearly as wide as they are long, dwarf the crazy scribble of small shorebird tracks found in profusion along the riverbank. The little birds scattered as the bigger animal strode past.

The big three-toed tracks belong to *Diatryma*, a 7-foot-tall (2 m) flightless bird weighing about 385 pounds (175 kg). It ranged across North America from New Mexico to Washington. The only known tracks in the entire world left by this giant bird were discovered in 2009, after a large landslide tumbled into Racehorse Creek, above the town of Kendall in the North Fork Nooksack River valley. About twenty *Diatryma* prints have been found in the rubble; most are single imprints, but there are a few closely spaced sequential tracks.

A Diatryma *footprint in the Chuckanut Formation, now on display in the Western Washington University Geology Department's corridors.* Inset: *Skin impression in the heel of the track.* —Courtesy of Tom Borst

A large sandstone slab with a solitary, nearly perfect *Diatryma* track greets you as you enter the Environmental Studies building's north entrance. The heel of the track left a dimpled impression of the giant bird's skin. There are many tracks of smaller birds on the slab. Very faint mammal tracks run right down the middle, from top to bottom.

The tracks found in the landslide are by far the world's oldest giant bird tracks. The track and its massive slab at the entrance were saved by the efforts of a dozen volunteers of the Big Bird Herd. They were concerned that commercial fossil hunters would try to illegally remove the valuable fossil from the public land where it was found or that it would be damaged in the process. Smaller bird tracks had already been destroyed by someone trying to chip them out of the slab. Also, the mudstone layer holding the track was beginning to delaminate from the sandstone beneath; once that happened the whole thing would quickly crumble. Big Bird Herd volunteers

Volunteers from the Big Bird Herd move the 1,300-pound (590 kg) rock slab with the Diatryma *track.*

designed and built a heavy wooden platform and packed it in pieces up the steep landslide slope from the nearest logging road. They heaved the 1,300-pound (590 kg) slab onto the platform, dragged it 30 yards (30 m) or so with small hand winches, and buried and disguised the whole thing to protect it until a logging helicopter could fly it to safety in July 2010.

In 1876 Edward Drinker Cope, one of the pioneers of North American paleontology, described fossil bones from a giant ground-bird found in Wyoming that he named *Diatryma gigantea*. The name *Diatryma* means "through a hole" in Greek, referring to perforations in some of the foot bones. Blood vessels and nerves passed through these openings. The *gigantea* part is self-evident! *Diatryma*

A reconstructed fossil skeleton of a 6.5-foot-tall (2 m) Diatryma *at the Smithsonian Museum of Natural History.* —Courtesy of Janine and Jim Eden

had a huge skull, 1.5 feet (0.5 m) long, and a massive bill to go along with its big feet. The wings were merely vestigial, and such flightless birds are often called groundbirds.

Very similar bones were described earlier in France and elsewhere in Europe. This bird was named *Gastornis*. In fact, footprints presumed to have been left by a *Gastornis* were found near Paris in 1859, but they have disappeared. They were found in rocks slightly younger than the Chuckanut. Perhaps *Gastornis* and *Diatryma* are the same bird, or at least in the same genus. There is disagreement in the paleontological world. *Diatryma* lived in North America during the early Eocene, but *Gastornis* stalked about in Europe from the mid-Paleocene to the middle Eocene. This family of large, flightless birds seems to have originated in Europe, then radiating to North America before the final connection between Eurasia and North America was severed as Pangaea rifted apart. *Diatryma* fossils have been found on Ellesmere Island in the Arctic.

It is easy to picture a big-billed, 7-foot-tall (2 m), 385-pound (175 kg) bird as a rapacious carnivore, and illustrations in books and on the Internet show a swiftly running bird with taloned feet clutching screaming animals in an eagle-like bill. Some scientific papers have hypothesized that the bird was the top terrestrial carnivore during the Eocene, evolving to fill the niche vacated by the recently extinct raptor dinosaurs. Some museums as well as university and other scientific websites continue to describe *Diatryma* as a large carnivore, yet others suggest the bird may have been a scavenger.

There have been very few scientific papers examining the bird from an anatomical perspective, and those that have do not reach consensus on whether or not the bird was a predator. Some argue that its giant bill was adapted for bone crushing, whereas others propose it was used for cutting tough vegetation. The bill is not hooked, as are those of carnivorous birds. *Diatryma*'s bill could have been adapted for cracking palm nuts or eating large fruits, as do some modern birds with large bills—toucans, for example. Some researchers point out that flightlessness is generally associated with a vegetable diet, because leaves take time and energy to digest—energy that would otherwise be needed for flight. *Diatryma*'s leg bones, with a long femur and short tarsus; its wide pelvis; and its short, broad toes, without anchors for the flexors of talons, all argue for a plodding gait, hardly conducive to fierce predatory behavior. *Hyracotherium* fossils are found with those of *Diatryma*, and with their tracks, but this only demonstrates similar habitats, not a predator-prey relationship.

The fossil tracks found in the Chuckanut Formation provide new evidence toward resolving the debate. The fleshy heel pad left the

Diatryma *has most often been depicted as a rapacious carnivore. This one has caught an unfortunate* Hyracotherium. —Courtesy of ©John Sibbick

Marlin Peterson's painting among the hallway displays shows Diatryma *as an herbivore eating fruit from a tall* Sabalites *palm.* —Courtesy of ©Marlin Peterson

deepest impressions in all the tracks, meaning the heel carried much of the bird's weight. The stride length between paired tracks is short and plantigrade, meaning the bird walked flat-footed with toes and heel both on the ground; if the bird had long claws, they would have left an impression. The toe impressions end in short and triangular toenails, not claws. The tracks do not mimic the widely separated, toe-walking gait of a running biped.

Why have no bones been found in the Chuckanut or correlative formations, such as the Huntingdon Formation along the border with British Columbia and the Swauk Formation in the Leavenworth-Peshastin area? The preservation of bones is a dicey thing, and only a tiny fraction of animal skeletons are fossilized. Fossils are particularly rare in sandstone, the predominate rock in the Chuckanut Formation. Oxygen readily permeates sandy deposits, which enhances the growth of microbes to break down remains. In river environments, flowing water scatters disarticulated bodies, so even if fossils are found, they are typically only a part of the animal. Furthermore, though river and swamp environments are rich in animals, they also include a lot of scavengers who eat the animal remains.

Just beyond the great sandstone slab, across from the elevator, is a fascinating display about a very controversial discovery. A large "fossil track" from the Green River in King County made quite a stir in northwest media when it was found in 1992. It was thought to be a *Diatryma* track, but it is clearly different from the tracks found in the Chuckanut Formation. Compare the fossil with the *Diatryma* track and see what you think.

Key differences include overall size, the angles between the toes, and the shape of the nail impressions. Also, the rock of the Green River impression is sandstone from the late Eocene Puget Group, putting the age of the specimen outside the known time range of *Diatryma*. These differences could be explained if the impression is the footprint of some other giant groundbird unknown from skeletal remains. Examine the Green River specimen very closely. The depressed areas do not look like they caused the sediment to deform. In particular, the "toe" and "heel pad" impressions cross-cut thin sedimentary layers, as if the sediment had already been lithified. Compare this impression with the other tracks, *Diatryma* and otherwise. You can see that they all deformed the soft mud those long-deceased animals stepped in. This characteristic, in particular, suggests that the Green River impression is a pseudofossil that merely resembles a bird track. If the impression is a footprint, it was not made by *Diatryma*. It was a wonderful coincidence that the Green River slab was already on display at Western Washington

University when the Chuckanut Formation tracks were found. This allowed for direct comparison.

The hallway museum displays other Chuckanut animal tracks and plant fossils. A painting by Marlin Peterson depicts snarling creodonts encircling a family of tapirs. There is a reproduction of a *Diatryma* skull, an exhibit about Bellingham's coal-mining past, and information about earthquakes. There are also mineral cases and two big glass display cases full of antique geologic tools. Exhibits can be found on the second, first, and ground floors, but be aware that in midweek classes will be in session, so please do be quiet. The museum is worth a visit every couple of years, because displays are updated or changed as new discoveries are made. And you can't beat the free admission!

Perhaps after reading this vignette you may have a nagging feeling that you've heard of this bird before. That's because *Diatryma* is a pop culture phenomenon. A French rock band bears the name. Terrible predatory *Diatryma*s inhabit video games. The bird is even pictured on postage stamps of Oman, Yemen, and Bahrain, which is curious, as no trace of the big bird has been found in these nations.

Challenge your rock identification skills on the first floor, at the opposite end of the building from the Diatryma *exhibit.*

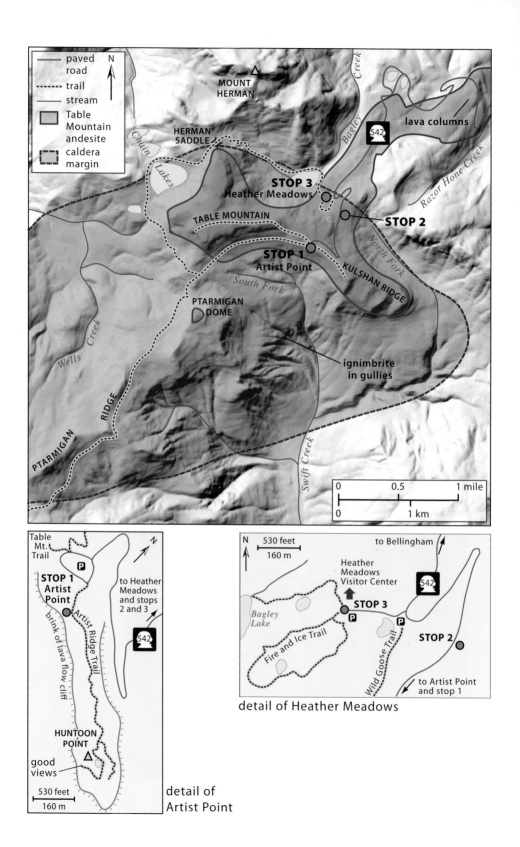

paved road
trail
stream
Table Mountain andesite
caldera margin

N

MOUNT HERMAN

Chain Lakes

HERMAN SADDLE

Bagley Creek

542

lava columns

Razor Hone Creek

STOP 3
Heather Meadows

TABLE MOUNTAIN

STOP 2

STOP 1
Artist Point

KULSHAN RIDGE

North Fork

South Fork

PTARMIGAN DOME

ignimbrite in gullies

Wells Creek

PTARMIGAN RIDGE

Swift Creek

0   0.5   1 mile
0   1 km

Table Mt. Trail

N

P

STOP 1
Artist Point

Artist Ridge Trail

brink of lava flow cliff

to Heather Meadows and stops 2 and 3

542

HUNTOON POINT

good views

530 feet
160 m

detail of Artist Point

N   530 feet
     160 m

to Bellingham

Heather Meadows Visitor Center

542

STOP 3

P

Bagley Lake

Fire and Ice Trail

P

Wild Goose Trail

STOP 2

to Artist Point and stop 1

detail of Heather Meadows

22

Calderas and Columns

ARTIST POINT AND HEATHER MEADOWS

Artist Point, elevation 5,040 feet (1,536 m), is a hub of hiking trails at the end of Washington 542. You may find walls of snow surrounding the parking lot, which usually isn't plowed out until midsummer, if by then. The Mount Baker Ski Area holds the world record for annual snowfall: 1,140 inches (2,896 cm) fell between July 1, 1998, and June 30, 1999. That equates to 95 feet (29 m) of uncompacted snow. That summer snowplows left vertical walls of snow 25 feet high (8 m) at the end of the highway. Even in an average year, between 50 and 60 feet (15 and 18 m) of snow falls in this region.

Find the Artist Ridge Trail at the southeast corner of the parking lot, across from the restrooms. The trail begins as an asphalt path that reaches a viewpoint (stop 1) in 100 yards (90 m). Mount

GETTING THERE: The stops in this vignette are in the Mount Baker–Snoqualmie National Forest, at the end of Washington 542 (Mount Baker Highway), a scenic highway running east up the Nooksack River from Bellingham. You may want to stop at the Glacier Visitor Center in the town of Glacier, about 33 miles (53 km) from I-5, to purchase a National Forest Recreation Day Pass or an annual Northwest Forest Pass. There is a chance you will be ticketed at the stops if you don't have a pass. On the way up the highway, feel free to indulge the temptation to pull over to examine the spectacular andesite columns along the road and to gawk at the outstanding scenery. We'll visit exposures of the same lava flow in Heather Meadows.

Stop 1 is at the end of the highway at the large Artist Point parking lot, 58 miles (93 km) from I-5, at the end of a series of tight switchbacks. Stop 1 is an easy walk on a paved path for about 100 yards (90 m). A gravel trail continues another 0.5 mile (0.8 km) along Kulshan Ridge if you wish an even better geologic overview. Stop 2 is a pullout about 1 mile (1.6 km) back down the road. Stop 3 is at the Heather Meadows Visitor Center, about 1 mile (1.6 km) down the road from stop 2.

This is a very popular area. In most years the road to Artist Point is snow-free for only two to three months, usually from late July to the end of September.

Shuksan (9,131 feet, 2,783 m), metamorphosed ocean-floor basalt sculpted by glaciers, is only about 4 miles (6.5 km) to the east. In clear weather the glacier-clad Mount Baker volcano, about 7.5 miles (12 km) to the southwest, dominates the view. At 10,781 feet (3,286 m), its cone is visible in the lowlands from Seattle to Vancouver, British Columbia. The oldest lavas known to have erupted from the volcano are only 40,000 years old, making Baker as young as Mount St. Helens. This active volcano constantly emits steam and other gases from Sherman Crater, which forms the prominent notch 1,000 feet (300 m) below the summit on Baker's south side. On clear, cold winter days, these volcanic emissions are visible from the populated lowlands, sometimes rising high into the frigid air. The smell of hydrogen sulfide, with its characteristic rotten-egg odor, sometimes wafts as far as Artist Point. Mount Baker is only the most recent stratovolcano to grow in this region. A dozen or more volcanoes have grown in the Mount Baker Volcanic Field over the past 1 million years or so. Magma compositions in the field range from basalt to rhyolite, and eruption styles reflect these compositions. Some volcanoes became high mountains; others erupted only once to form small cinder cones. One was a caldera.

Repeated glacial advances stripped away much of these older volcanic edifices. Only bits and pieces were left behind for geologists to discover and map as they tried to understand the area's past record of volcanism. This vignette investigates rocks and landforms related to a pair of volcanic episodes that predate the growth of Baker's cone: the Kulshan caldera and the Table Mountain andesite.

At stop 1 you are perched on an eroded remnant of Table Mountain andesite, a collection of gray lava flows that erupted 300,000 years ago. The lava source was on Ptarmigan Ridge, about 1.5 miles (2.4 km) west of Artist Point. These flows form the aptly named Table Mountain to the west and the steep cliffs below you; they extend east to the far end of Kulshan Ridge at Huntoon Point. Prior to erosion, the Table Mountain andesite continued an unknown distance south from Artist Point into the valley of Swift Creek. The lava columns you drove past on the way up the highway are part of these lava flows.

The lava overlies older volcanic rock visible in the deep valley of Swift Creek, to the south. The white to pale-gray rocks about 1,000 feet (300 m) below you are part of the ash and debris that filled the Kulshan caldera during its catastrophic eruption and collapse 1.15 million years ago, the oldest volcanism associated with the Mount Baker Volcanic Field.

Mount Baker from Artist Point. The active Sherman Crater lies in the deep notch left of the summit. The flat summit is the ice-filled Carmelo Crater. Older lava flows lie in the foreground.

Calderas like the Kulshan are a special type of volcano. A well-known example in the Cascades is the 7,000-year-old Crater Lake caldera in southern Oregon. The caldera left behind by the Kulshan eruption is similar in size to Crater Lake. There was a lake here, too, but glacial erosion destroyed the walls of the caldera, draining the lake long ago. The huge caldera at Yellowstone may come to mind. It dwarfs any caldera in the Cascades, having erupted about 240 cubic miles (1,000 km^3) of ash and debris in a single eruption.

The Kulshan caldera measures 3 by 5 miles (5 by 8 km), elongated along an east-west axis. It resulted from the eruption of about 12 cubic miles (50 km^3) of silica-rich rhyodacite magma, with a chemical composition on the cusp between dacite and rhyolite. The ash eruption was probably continuous, perhaps lasting a few days at most. Compare that with what is characteristically described as the catastrophic eruption of Mount St. Helens on May 18, 1980, which erupted a piddling 0.12 cubic mile (0.5 km^3) of magma. The northern portion of Washington was in the grip of an ice age at the time. The thickness of ice at the site of the eruption is not known, but it is evident from the texture of the ash filling the caldera that the erupting magma mixed with a lot of water, and the age of the eruption coincides with a period of glaciation that extended to at least the south end of Puget Sound.

Calderas do not always develop at sites of older volcanism. Crater Lake caldera formed at the lofty Mount Mazama volcano, but there is only scant geologic evidence that any volcanic activity preceded the Kulshan caldera. Before calderas form, large volumes of silica-rich magma gradually accumulate at shallow levels in the crust. As the magma rises into the lower-pressure environment close to the surface, dissolved gases, principally water, carbon dioxide, and sulfur dioxide, begin to come out of solution to form bubbles, but the bubbles can't escape from the viscous magma. The silica-rich magma is hot and of low density and may rise to within 0.5 mile (1 km) or so of the surface. The upward pressure from the growing accumulation of gas in the magma causes the crust to bulge upward. This may cause less than 100 feet (30 m) or so of doming, but it is enough to crack the crust, typically along an arcuate ring fracture that develops near the perimeter of the bulge, as well as in the center of the bulging crust. The fractures form pathways for the pressurized magma to explosively erupt as ash and pumice.

The difference between a caldera and other volcanoes is the large supply of silica-rich magma near the surface, the doming at the surface, and the fracturing in the crust above the magma. Ordinarily, as the volume of magma decreases in an eruption, so does the gas pressure in the magma. Eventually the pressure is reduced to the point that the eruption ends. A caldera eruption, however, is more like a runaway train. The magma reservoir decreases in volume as huge amounts of hot, glowing ash and pumice rapidly erupt from the ring fracture. The overlying 0.5-mile-thick (0.8 km) crust that roofs the magma chamber subsides rapidly into the emptying reservoir along the ring fracture as it loses support from the underlying magma. The roof may subside as a coherent plate in a piston-like way, or in individual pieces. The collapsing roof exerts pressure on the magma, so the eruption continues until so much gas-charged magma has erupted that there is insufficient pressure remaining in the chamber to drive the eruption. As the roof subsides, landslides may peel off the steepening caldera walls, sliding into the developing caldera. The subsiding roof forms the floor of the caldera, and the ring fracture—now a fault because movement has occurred along

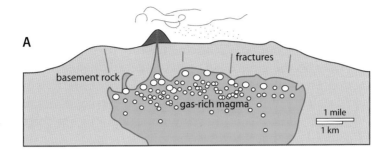

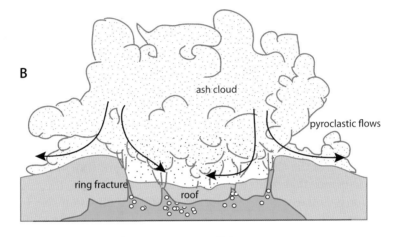

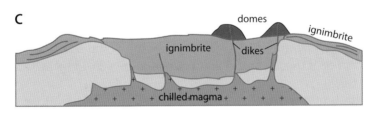

Caldera formation. (A) Large volumes of gassy silica-rich magma accumulate near the surface. Fractures develop in the crust above the magma as the surface is domed upward; the fractures coalesce and outline the dome, thus the term ring fracture. Some magma may erupt from the fractures to form volcanoes. (B) Pressure builds and more fractures develop until magma erupts from the ring fracture in huge columns of ash and pumice. The roof of the magma chamber collapses as magma is emptied. Collapsing columns give rise to pyroclastic flows, which cover the landscape. Ash, pumice, and rock fragments fill the subsiding caldera to form ignimbrite. (C) After the eruption, pyroclastic flow deposits cover the basement rock. The caldera is filled with a great thickness of ignimbrite. Dikes may intrude the caldera to erupt as domes. The magma chamber cools and becomes solid rock.

it—forms the walls of a vast crater. In the case of Kulshan, its crater dwarfs anything seen on the stratovolcanoes Rainier and Baker.

Ash and pumice pour out of the ring fracture to form pyroclastic flows that may fill the great bowl of the caldera as it subsides. The resulting deposit is called ignimbrite, a type of pyroclastic flow high in silica and rich in pumice. Ignimbrites also contain solid rock fragments torn from the walls of the eruption caldera by the forceful rush of fractured magma. The towering eruption cloud rising above a subsiding caldera can also collapse from its own mass and spread outward over the countryside as ignimbrite flows, burying everything for tens of miles (kilometers) around under sheets of searing ash, pumice, and rock fragments. Pumice blocks and ash fragments in the ignimbrite may compact as erupted material continues to accumulate, and if they are hot enough, they weld together as a dense volcanic rock called welded tuff, which can take years to cool completely. Erosion has revealed more than 3,300 feet (1,000 m) of ignimbrite in the Kulshan caldera. Since the caldera floor is nowhere exposed, the ignimbrite could be even thicker.

In the case of the Kulshan caldera, the pyroclastic flows blanketed glacial ice; once that melted away due to climate change, the deposits similarly collapsed and were carried off by meltwater. Ash from the eruption traveled great distances before settling to the surface. It, too, blanketed an icy landscape, so it was washed away when the glaciers melted. The only recognized remnants of what must have been a thick blanket are found at five locations near Auburn and Hoodsport, Washington, 120 miles (190 km) to the south, beyond the terminus of the glacier. The ash deposits are 4 to 12 inches (10 to 30 cm) thick.

By the time the Kulshan eruption ended, the caldera floor had subsided at least 3,300 feet (1,000 m), and a vast hole had been blasted through the ice. The caldera was largely filled with ignimbrite. However, a semicircular wall of bedrock, the rim of which was also coated with ignimbrite, marked the margins of the subsided crust. In no time a lake began to fill the depression. It would have been turbid with ash washing into it off the surrounding icy landscape. The lake likely steamed heavily, and large ice blocks floating in the dirty water jostled each other. Magma, now less gassy following the great eruption, occasionally extruded upward through the lake to form lava domes. Alpine glaciers, streams, and several later continental glaciations further reamed out the caldera, breaching its rim and draining away the lake. In places, sediment from the floor of the lake is preserved above the caldera-filling ignimbrite.

At Artist Point you are standing near the northeast margin of the caldera. The 3,300-foot-thick (1,000 m) whitish ignimbrite below and to the south has been eroded into knife-sharp ribs and deep gullies. The southern bedrock rim of the caldera is the ridge to the south and west forming the skyline between you and Mount Baker. The southern branch of Swift Creek, flowing to Baker Lake 8 miles (13 km) to the south, has breached the southeast margin of the caldera. Between the stream and repeated, profound glacial erosion, the eastern portion of the caldera's interior has been gutted. Consequently, Kulshan caldera is beautifully laid out in cross section for geologic study, although reaching the interior is a strenuous off-trail adventure in forbidding, dangerous terrain.

Young calderas are uncommon in the Cascade Volcanic Arc. Other than Kulshan and Crater Lake, the only other known Pleistocene-age caldera is the 600,000-year-old Rockland caldera, largely buried beneath the younger rocks of Lassen Peak in northern California.

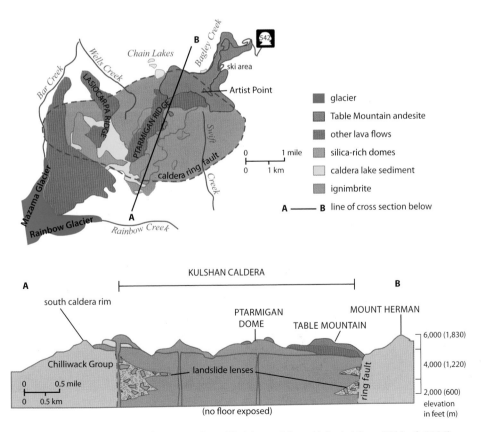

Simplified geologic map and cross section of Kulshan caldera. (Adapted from Hildreth 1996).

There are several much older calderas in the Washington Cascades. Most of them, however, are millions of years old and hard to recognize as calderas. An exception is the Hannegan caldera, just beyond Mount Shuksan on the eastern skyline; it is only 3.72 million years old and beautifully cross-sectioned by erosion. Its fill of welded ignimbrite forms the south face of Hannegan Peak.

A wide gravel trail continues east along the crest of Kulshan Ridge. It is a nearly level 0.5-mile-long (0.8 km) trail to the east end of Kulshan Ridge, where there are more fine views into the gutted interior of the caldera. For great views into Kulshan caldera, follow the trail beyond a small tarn, and then pick a route on boot-beaten tracks past a couple of possibly dry ponds on the eastern side of Huntoon Point, the high point of the ridge. Walk to the southern margin of the ridge. At the top of a 200-foot-high (60 m) lava cliff, elevation 5,200 feet (1,585 m), the buff to light-gray ignimbrite is easily seen in the yawning valley of the U-shaped, glacially carved Swift Creek. Near the headwaters to your right, the ignimbrite is, in places, very thinly capped with the lowest flow of the Table Mountain andesite.

Hike back to the parking lot. The Table Mountain andesite that composes the ridge was scoured, scraped, and smoothed by grit carried in glacial ice. Look for glacial striations oriented north to south on flat or slightly rounded lava knobs. Back at the stop 1 viewpoint take the left fork to a paved, stone-walled viewpoint less than 20 feet (6 m) away. It is directly above the southeast corner of the parking lot. This is a good place to see glacial striations and enjoy the view of Mount Baker.

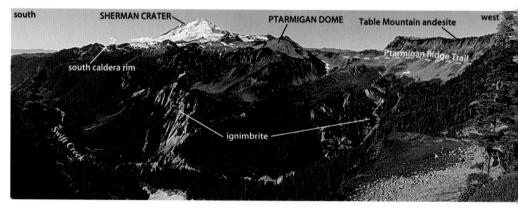

From the viewpoint south of Huntoon Point, the southern caldera rim forms the southern skyline. Lava of Ptarmigan Dome, erupted shortly after the caldera-forming eruption, is capped with Table Mountain andesite.

On the way to stop 2 you pass the Lake Ann Trailhead, which lies on the northern part of the ring fault of the Kulshan caldera. The fault continues eastward down the valley of Swift Creek. You can't miss the curved, horizontal columns on the east side of the highway, the right side as you head downhill. Carefully park in a small turnout across the road from the spectacular outcrop. This is stop 2.

Exposed in the outcrop is Table Mountain andesite that flowed against the bedrock of the steep valley wall. The lava once extended northwest across Bagley Creek to the slopes of Mount Herman, but it has been eroded down to the level of the Heather Meadows picnic area. If you walk down the road 50 feet (15 m) or so, you can look up the slope to see the very steep margin of the lava-bedrock contact. The greenish bedrock itself is exposed another 100 yards (90 m) down the road. This is the 255-million-year-old (Permian age) Chilliwack Group, a mix of submarine lava and lithified seafloor sediment that was accreted to the margin of North America (see "E Pluribus Unum: Assembling Washington" in the introduction). These green rocks are relatively fracture-free compared to the lava, and harder.

Andesite starts solidifying at around 2,000°F (1,100°C). Lava contracts as it cools, so fractures, or joints, develop within the rock to accommodate the reduction in rock volume. Joints grow as a cooling front, at about 1,380°F (750°C) or so, progresses into the lava flow from its outer surfaces, growing incrementally into the lava perpendicular to the margin of the cooling front. They start out at right angles to the lava surface, which has been chilled where it contacts the atmosphere or the cold rocks of Earth's surface. As the joints grow they intersect and surround columns of hot rock, and thus they are known as polygonal or columnar joints. The joints typically intersect at an angle of around 120°.

If the lava is bathed in water or is in contact with ice, cooling is more rapid and joints develop closer together, defining narrower columns. Slower cooling produces wider columns. Cooling joints can't form if the lava is still advancing since the stress of motion deforms or destroys them. Columns are often associated with basalt lava (vignette 5) but can be found in lavas or welded tuffs of any composition (see vignette 2 for another example in andesite). Textbook columns are hexagonal, but six sides is only the idealized perfect-world form. Real-world columns have as few as three and as many as twelve sides.

The columns at stop 2 radiate away from each other and swoop upward and outward toward the road. That is because the lava was cooling inward, toward the road from the colder rocks of the valley wall (remember, the interior of the lava has eroded away; you are standing where the lava used to be). Because the cooling front moved into a three-dimensional space (the inside of the lava flow) from an irregular two-dimensional contact (between the lava and

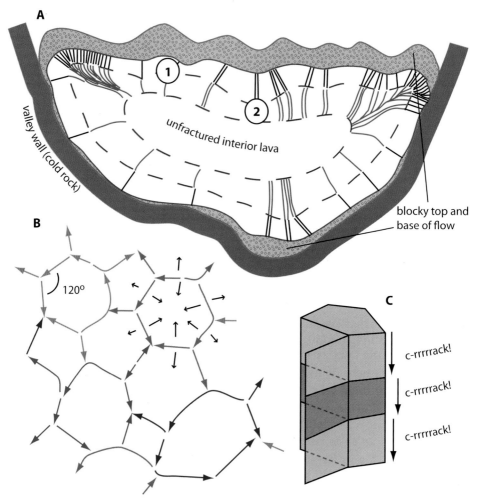

How columnar jointing develops. (A) A cross section through a lava flow in a steep-sided valley. The cooling front (dashed lines) migrates inward. Cooling joints (solid lines) move inward at a right angle to the cooling front. At time 1, the joints (black) have migrated a short way into the flow. As cooling progresses, so do the joints (red). Some merge to create radiating fan shapes. (B) View parallel to the surface of the cooling front. Contraction during cooling produces tension within the lava (black arrows illustrate this in a selected area). Joints form perpendicular to the tension and open in a variety of directions (colored arrows). Joints continue forming until they intersect other joints, typically at or near 120°. (C) Cooling joints progress into a lava flow in increments separated by minutes to days, depending on the cooling rate. (Part C is modified from Aydin and DeGraff 1988.)

Diverging cooling fronts (arrows) spread from the cold bedrock into the Table Mountain andesite lava to form curved columns.

the valley wall), it spread radially into the lava; the joints following the front radiated as well, forming this artful spray of rock columns.

Continue down the road to the Heather Meadows Visitor Center (stop 3) and park at the end of the road. The visitor center is built of stacked "logs" of andesite. Beautiful, classic cross sections through the andesite columns are exposed on the glacially polished rock slab just beyond the low stone wall at the west end of the parking lot. Walk onto this slab. We are near the bottom of the lowest flow in the stack of Table Mountain andesite. These columns grew upward into the flow from below. The rasp of glaciers sanded the andesite down to this level. Fine examples of glacially smoothed columnar andesite line the road lower in the meadows.

The self-guided interpretive Fire and Ice Trail begins in front of the visitor center. It is paved and free of barriers for the first 0.2 mile (0.3 km). The rest of the trail is fairly level gravel with a few steps; it loops back to the visitor center in another 0.3 mile (0.5 km).

These columns along the Fire and Ice Trail are exposed in cross section thanks to glacial scouring. Moss growing in the fractures accentuates the outlines of the columns, which have a variable number of sides.

I recommend a counterclockwise stroll. Along the loop you'll find a sign labeled "A valley in the sky," which interprets the erosional history of Table Mountain as a classic example of reversed topography. The sign says that the lava once lay within a deep valley, and that its prominent height is the result of streams eroding away the "soft" rock of the valley walls that once confined the lava. When a lava flow covers a valley bottom, the stream that once ran along the valley floor is displaced, but it often reestablishes itself as two streams, one on each margin of the lava flow. These streams erode the contact between the lava and the valley wall; eventually the lava is left as a ridge between two younger valleys. This phenomenon is known as topographic reversal and is well demonstrated in places.

The "valley in the sky" interpretation, however, predates the discovery of the Kulshan caldera in 1996. It seems very unlikely that a stream beginning at Herman Saddle, only 1 mile (1.6 km) from this sign, could have produced this profound topographic change. A different interpretation can be made for Table Mountain's morphology.

A stack of four viscous lavas, each of which piled up for 300 to 400 feet (90 to 120 m), were erupted one after the other about 300,000 years ago to become the Table Mountain andesite. They erupted from a volcano, now eroded away, on Ptarmigan Ridge, 1.8 miles (2.9 km) southwest of Artist Point; today only a dike remains to indicate the eruption site. Remnants of these lava flows make up the whole mass of Table Mountain and its eastern extension, Kulshan Ridge. The oldest exposed flow in the stack covers the floor of Bagley Creek's little valley. The lava flowed out over the caldera-filling ignimbrite and lake sediment. Successive flows covered or lapped against earlier ones and banked against the high topographic wall forming Kulshan caldera's north margin, where they piled up to make the imposing thickness that would become Table Mountain. To the south, steep-fronted lava lobes oozed out over the caldera-filling ignimbrite and lake sediments. On their southeast and west margins, at least the lowest lava was quenched against ice that filled a portion of the caldera; the eastern glacier was probably coming from what would become Mount Shuksan. Some of the lava overtopped the caldera rim and headed northeast an unknown distance toward the North Fork Nooksack River; you saw this lava along the highway on your way up.

The lava of Table Mountain, however, lies entirely within the bounding walls of the caldera. This is important: the lava flows that were later eroded into Table Mountain were not confined in the valley necessary for the "valley in the sky" interpretation. There are two other crucial pieces of evidence that refute the interpretation. First, as you saw at stop 2, the andesite is fractured and far more susceptible to erosion than the older and harder metamorphosed Chilliwack Group basalt of Mount Herman, north of Table Mountain. And second, at least three continental ice sheets inundated the northern Cascades (vignette 11) after the eruption of the Table Mountain andesite. The steep-sided, flat-floored, U-shaped valley eroded into the north side of Table Mountain—and holding Bagley Lake—is a classic glacial cirque, probably occupied by different alpine glaciers during different glacial periods. If we consider these facts—the location of the Kulshan caldera and the bedrock wall on its northern margin, the greater susceptibility of the Table Mountain andesite to erosion relative to the Chilliwack Group rocks, and the fact that multiple glaciations have been at work here—it's more likely that glacial erosion was the principal process that removed the lava on all sides of Table Mountain, not stream erosion.

The aptly named Table Mountain consists of 1,500 feet (450 m) of andesite rising above the floor of Bagley Creek. The northern ring fault of the Kulshan caldera, buried by lava in the valley, passes under Herman Saddle, the pass on the right. —Courtesy of Bob Mooers

At the "valley in the sky" sign, you are standing above the northern part of the ring fault of the Kulshan caldera, buried beneath the lava covering the valley floor. The fault runs westward up the valley and through Herman Saddle, separating Table Mountain andesite inside the caldera from the metamorphosed Chilliwack Group seafloor basalt that composes Mount Herman, across the valley on the caldera's northern margin.

As you walk the loop, note the scattered boulders of greenish breccia lying in the heather on the lava. The rock in the boulders is a jigsaw of broken, angular, light-colored fragments in a green matrix. These boulders are glacial erratics from the Chilliwack Group. The subsided crust that once lay above the developing magma chamber, on this northern edge of the caldera, is almost surely composed of Chilliwack Group rocks.

As you stand above the ring fault, contemplate for a moment the enormity of change that occurred here during the caldera's collapse.

Chilliwack rocks on the south side of the ring fault that were once continuous with the rocks of Mount Herman subsided nearly instantaneously; they now lie as much as 3,300 feet (1,000 m) below you, buried by caldera-filling ignimbrite. Since then, glacial erosion, lava flows, more glaciation, and stream erosion have altered the landscape, as they will continue to do unforeseeably into the future.

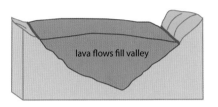
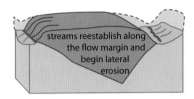

lava flows fill valley

streams reestablish along the flow margin and begin lateral erosion

Topographic reversal interpretation for Table Mountain.

1. Table Mountain lava flows

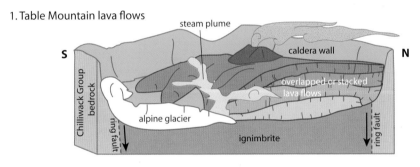

steam plume

S

N

Chilliwack Group bedrock

ring fault

caldera wall

overlapped or stacked lava flows

alpine glacier

ignimbrite

ring fault

2. Repeated glaciations

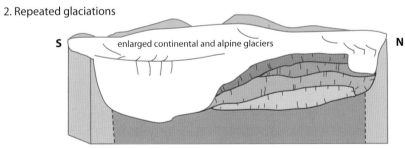

S

N

enlarged continental and alpine glaciers

3. Today

S

N

South Fork Swift Creek

STOP 1

Bagley Creek

STOP 3

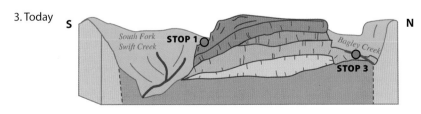

Glacial erosion interpretation.

GPS POSITIONS FOR SELECTED WAYPOINTS

This table provides UTM (WGS 84, zone 10) and latitude/longitude positions for stops and other important waypoints in this book. Positions were obtained by using a handheld GPS receiver or Google Earth™ (if the position was clearly visible in that program). UTM positions are rounded to nearest 10 meters.

Stop/Location or Geologic Feature	UTM Easting WGS 84 Zone 10	UTM Northing WGS 84 Zone 10	Latitude N Degree and Decimal Minute	Longitude W Degree and Decimal Minute	Elevation in feet (approximate)
1. Bridge of the Gods and the Bonneville Landslide					
1. Bridge of the Gods mural	585760	5057090	45°39.70' N	121°53.95' W	100'
2. Cascade Locks, Thunder Island	585980	5057800	45°40.11' N	121°53.77' W	80'
3. Wauna Viewpoint parking	583790	5054640	45°38.43' N	121°55.49' W	
Wauna Viewpoint	583540	5054290	45°38.23' N	121°55.69' W	1,040'
4. hummocks island pullout	586030	5059740	45°41.15' N	121°53.72' W	95'
2. Beacon Rock					
1. boat launch	576320	5052520	45°37.32' N	122°1.26' W	30'
2. Beacon Rock parking lot	576280	5053290	45°37.72' N	122°1.21' W	270'
3. Cape Disappointment					
1. Beards Hollow Overlook	418013	5128985	46°18.57' N	124°3.88' W	185'
O'Donell's Island	417610	5128590	46°18.35' N	124°4.20' W	44'
2. McKenzie Head	418100	5126320	46°17.13' N	124°3.79' W	51'
3. pillow lavas	417200	5127470	46°17.75' N	124°4.51' W	10'
4. Cape Disappointment Trail	418800	5125500	46°16.69' N	124°3.24' W	185'
4. The Sunken Forest of Willapa Bay					
1. Sandy Point	427060	5161730	46°36.31' N	123°57.15' W	10'
5. The Columbia River Basalt in the Chehalis River					
1. basalt along the River Road	486150	5164620	46°38.10' N	123°10.85' W	250'
2. Rainbow Falls	482230	5164160	46°37.85' N	123°13.93' W	260'

6. The Mima Mounds

1. interpretive center dome	496300	5194560	123°2.92' W	46°54.27' N	245'
2. Mima mound cutaway	495860	5192740	123°3.27' W	46°53.29' N	210'

7. Mount St. Helens National Volcanic Monument

1. roadside pullout at toe of lava flow	558320	5100500	122°14.76' W	46°3.34' N	575'
2. Trail of Two Forests	560840	5105380	122°12.77' W	46°5.96' N	1,900'
3. Ape Cave main entrance	560940	5106530	122°12.69' W	46°6.58' N	2,090'
Cave Basalt lava flow trail (use parenthetical comments if descending from upper entrance)					
stream crossing	560997	510671	122°12.640' W	46°6.654' N	2,070'
tumulus	561045	5106763	122°12.603' W	46°6.703' N	2,100'
shatter ring	560914	5107055	122°12.702' W	46°6.861' N	2,145'
inflation holes	560633	5107317	122°12.918' W	46°7.004' N	2,230'
(regain) lava surface	560527	5107411	122°13.000' W	46°7.056' N	2,295'
(leave) lava surface	560477	5107538	122°13.037' W	46°7.124' N	2,335'
(regain) lava surface	560384	5107769	122°13.108' W	46°7.250' N	2,400'
above the flow	560401	5107878	122°13.094' W	46°7.308' N	2,430'
alluvium in old roadcut	560414	5107959	122°13.083' W	46°7.352' N	2,445'
(leave) lava surface	560415	5108072	122°13.082' W	46°7.413' N	2,445'
Ape Cave upper entrance	560480	5108150	122°13.03' W	46°7.46' N	2,460'

8. Johnston Ridge

1. Johnston Ridge Observatory	560350	5124920	122°13.00' W	46°16.51' N	4,200'
2. Devils Elbow	562490	5124230	122°11.34' W	46°16.12' N	4,140'
3. Harrys Ridge	563640	5124970	122°10.44' W	46°16.52' N	4,800'

9. Sunrise, Mount Rainier National Park

1. Emmons Vista	603480	5196300	121°38.46' W	46°54.73' N	6,400'
2. pullout at lava columns	608310	5195880	121°34.67' W	46°54.45' N	4,670'
3. Osceola Mudflow deposit	605956	5193668	121°36.55' W	46°53.29' N	3,840'

Stop/Location or Geologic Feature	UTM Easting WGS 84 Zone 10	UTM Northing WGS 84 Zone 10	Latitude N Degree and Decimal Minute	Longitude W Degree and Decimal Minute	Elevation in feet (approximate)
10. Nisqually Glacier					
trailhead	596050	5182040	46°47.10' N	121°44.50' W	5,400'
1. Nisqually Vista	595660	5182260	46°47.22' N	121°44.81' W	5,250'
2. Glacier Vista	596490	5183850	46°48.08' N	121°44.10' W	6,340'
11. Whidbey Island					
1. Chuckanut erratic	535930	5313610	47°58.49' N	122°31.12' W	10'
2. brush-choked gully	535150	5313210	47°58.27' N	122°31.75' W	10'
3. southern point, Double Bluff	533860	5312770	47°58.04' N	122°32.78' W	10'
4. northern point, Double Bluff	533450	5313350	47°58.36' N	122°33.10' W	10'
12. Glacial Erratics of the Puget Lowland					
1. Donovan erratic	538660	5396150	48°43.03' N	122°28.47' W	740'
2. Coupeville's Big Rock	523270	5339730	48°12.63' N	122°41.21' W	120'
3. Waterman erratic	540020	5323290	48°3.69' N	122°27.79' W	230'
4. Lake Stevens erratic	565950	5316340	47°59.82' N	122°6.95' W	300'
5. Wedgwood erratic	552740	5281050	47°40.84' N	122°17.84' W	300'
6. Lake Lawrence erratic	532110	5190250	46°51.90' N	122°34.74' W	480'
13. Save Snoqualmie Falls!					
1. Peregrine Viewpoint	587360	5266240	47°32.62' N	121°50.34' W	430'
2. Lower Falls Viewpoint	587210	5266180	47°32.60' N	121°50.47' W	125'
3. volcanic rocks near Fish Hatchery Road	586750	5207090	47°33.08' N	121°50.82' W	120'
14. A Tour of Downtown Seattle Building Stone					
begin tour at Pioneer Square	550080	5272280	47°36.12' N	122°20.02' W	30'
end at Grand Hyatt Hotel	550070	5273480	47°36.76' N	122°20.02' W	140'
15. The Folded Rocks at Beach 4					
1. angular unconformity	395710	5278370	47°39.01' N	124°23.32' W	10'
2. chevron folds	395650	5278540	47°39.08' N	124°23.35' W	10'

Location	Easting	Northing	Latitude	Longitude	Elevation
16. Hurricane Ridge					
1. lava pillows	471230	5315970	47°59.78' N	123°23.14' W	3,160'
pillow contact over breccia	471550	5315970	47°59.78' N	123°22.88' W	3,080'
2. Hurricane Ridge Fault	464420	5313852	47°58.62' N	123°28.60' W	4,595'
3. Hurricane Ridge Visitor Center	462810	5312970	47°58.14' N	123°29.90' W	5,230'
17. Dungeness Spit					
1. bluffs at the base of the spit	485980	5332480	48°8.72' N	123°11.31' W	10'
2. New Dungeness Lighthouse	491810	5336510	48°10.91' N	123°6.63' W	5'
18. Skagit Gorge					
1. Gorge Creek Overlook	631700	5395650	48°42.00' N	121°12.61' W	1,095'
2. Diablo Lake Overlook	640000	5396940	48°42.59' N	121°5.82' W	1,700'
19. Washington Pass					
1. Washington Pass overlook	673700	5377480	48°31.58' N	120°38.84' W	5,630'
2. Blue Lake	671920	5375330	48°30.45' N	120°40.33' W	6,320'
20. Larrabee State Park					
1. gravel beach	537340	5388960	48°39.15' N	122°29.58' W	5'
2. Clayton Beach	538150	5387590	48°38.41' N	122°28.93' W	5'
21. *Diatryma*					
carpool area, parking lot C	537773	5397657	48°43.85' N	122°29.18' W	
ground floor entrance, ES Bldg.	537804	5397936	48°43.99' N	122°29.15' W	
22. Artist Point and Heather Meadows					
1. viewpoint near the parking lot	595960	5411100	48°50.73' N	121°41.52' W	5,072'
viewpoint east of Huntoon Point	596370	5410730	48°50.52' N	121°41.19' W	5,225'
2. pullout at andesite columns	596736	5412066	48°51.24'N	121°40.88'W	4,600'
3. Heather Meadows Visitor Center	596390	5412120	48°51.27' N	121°41.16' W	4,410'

Glossary

aa. A jagged, spiny basalt lava flow. *See also* pahoehoe.

accreted terrane. A portion of one plate that breaks off and becomes attached to an adjacent plate.

alluvium. Unconsolidated sediment deposited by rivers and streams.

amphibole. A group of dark-colored mafic minerals with alternating 60° and 120° angles between their four crystal faces. Some amphibole minerals form during metamorphism. *See also* hornblende.

andesite. A fine-grained volcanic rock, either lava or tephra, with a silica content intermediate between basalt and rhyolite. It is composed mainly of plagioclase with other minerals, such as hornblende, pyroxene, or olivine. Andesite is the volcanic equivalent of diorite.

arc. *See* volcanic arc.

argillite. A slightly metamorphosed, fine-grained rock. The protolith is shale or mudstone.

ash. Tephra fragments less than 0.08 inch (2 mm) across.

asthenosphere. A weak layer of hot, ductile rock in the upper mantle, below the lithosphere.

augite. A dark-green to black variety of clinopyroxene commonly found in Cascade Range volcanic rocks.

basalt. A mafic, dark, fine-grained volcanic rock, either lava or tephra, containing less silica than andesite. It is composed mainly of plagioclase with other minerals, principally olivine and pyroxene.

basaltic andesite. Andesite that is close to basalt in composition but has silica content just into the andesite range. Minerals are chiefly plagioclase and pyroxene. Olivine may be present.

basement. Undifferentiated bedrock beneath rocks on the surface.

batholith. An igneous intrusion with a surface exposure exceeding 40 square miles (100 km^2). Batholiths may consist of one or more plutons.

bed. The smallest sedimentary unit that is distinguishable from those above and below it.

bedding. The layered structure in sedimentary rocks caused by variations in sediment size and composition.

bedrock. The solid rock beneath unconsolidated sediment.

bioturbation. Soil disturbance by burrowing animals or plant roots.

bomb. Tephra fragments more than 2.5 inches (64 mm) across.

boulder. A rock fragment greater than 10 inches (25 cm) across.

breccia. A sedimentary or igneous rock consisting of angular fragments of other rocks.

brittle. Said of rocks that fracture under stress. *See also* ductile.

calcareous. Containing calcium carbonate.

calcite. An abundant mineral composed of calcium carbonate. It is the principal component in limestone, travertine, marble, and onyx marble.

caldera. A volcanic depression, often filled with ash and rock fragments, formed by the subsidence of the surface during the eruption of large volumes of magma.

Cascade Volcanic Arc. The 875-mile-long (1,400 km) range of volcanoes running the length of the Cascadia Subduction Zone from near Mount Meager in southern British Columbia to the Lassen volcanic field in northern California. Eruptions have occurred at 2,339 distinct vents during the Quaternary.

Cascadia Subduction Zone. The seismically active portion of the North American Plate eastward of its contact with the Juan de Fuca Plate to the Cascade Volcanic Arc, including the thrust fault separating the plates.

chert. A hard, dense, microcrystalline sedimentary rock consisting principally of quartz crystals.

cinder cone. A pile of scoria (cinders) that was erupted into the air and then accumulated around the vent of a short-lived volcano.

cirque. A glacially eroded amphitheater on the side of a mountain or at the head of a valley. A glacier may or may not persist in a cirque.

clast. An individual fragment of rock in sedimentary or volcanic rocks and deposits.

clay. A rock fragment less than 0.00016 inch (0.004 mm) across. Also a deposit consisting of clay-sized grains, and a family of minerals rich in water, aluminum, and silica with sheetlike crystal structures.

cleavage. Planes of weakness along which minerals tend to fracture.

clinopyroxene. A dark, mafic silicate mineral in the pyroxene family containing calcium or sodium. The boxy crystals have right-angle cleavage with smooth edges. *See also* augite.

Coast Plutonic Complex. The huge mass of intrusive and metamorphic rocks found in the northernmost Cascades, the British Columbia Coast Mountains, and the Kluane Range in southwestern Yukon. It is widely considered the world's largest belt of plutonic rocks.

cobble. A rock fragment between 2.5 and 10 inches (6 and 25 cm) across.

columnar joints. Joints formed in cooling lava as it contracts. They outline columns of rock.

conglomerate. A sedimentary rock made of more or less rounded grains of other rocks.

convergent plate boundary. A place in the lithosphere where two tectonic plates move toward each other. *See also* subduction zone.

cooling front. A migrating temperature boundary between hotter and cooler rock.

cooling joint. Fractures generated as a rock cools and contracts.

Cordilleran Ice Sheet. A sheet of glacial ice composed of conjoined glaciers flowing outward from the Coast Mountains in Canada during the Pleistocene epoch. *Cordillera*, Spanish for "mountain chain," refers to the mountains along the Pacific coast of North America.

country rock. Rock that is intruded by magma or mineral veins.

creodont. A member of an extinct order of carnivorous mammals that lived from the Paleocene to the Miocene epochs. Creodonts shared a common ancestor with the order Carnivora.

crust. The igneous, sedimentary, and metamorphic rocks composing Earth's outermost layer, which is up to 60 miles (100 km) thick.

crystal. A solid element or compound in which the atoms are arranged in an orderly and repeating pattern bounded by planar surfaces. Most minerals take a crystalline form if they can grow in a fluid medium, such as magma or water.

dacite. A dark, fine-grained, silica-rich volcanic rock, either lava or tephra. It is composed mainly of plagioclase with other minerals, such as quartz, hornblende, pyroxene, and biotite, and is the volcanic equivalent of granodiorite.

debris flow. A viscous slurry of sediment and water that moves rapidly down a slope or river valley. Debris flows are dense enough to float boulders. *See also* lahar.

deformation. The folding, faulting, shearing, compressing, or extending of rocks.

delta. The submerged fan of sediment at the mouth of a stream flowing into a lake or sea.

dike. A tabular igneous intrusion that crosscuts other rock structures, such as bedding planes or other intrusions. *See also* sill.

diorite. A gray to dark-gray, medium- to coarse-grained intrusive igneous rock composed principally of plagioclase, biotite or hornblende, with or without pyroxene. It contains less silica than granodiorite and is the plutonic equivalent of andesite.

dip. The downward inclination of a tilted rock bed.

discharge. Water volume flowing past a given point per unit of time. Commonly expressed in cubic meters or cubic feet per second.

divergent plate boundary. A place in the lithosphere where two tectonic plates move away from each other. *See also* spreading center.

drift. Material deposited either directly from ice or from meltwater issuing directly from ice. *See also* glaciomarine drift.

ductile. Said of rocks that are able to sustain a certain amount of bending, stretching, or compressing before fracturing under stress. *See also* brittle.

edifice. The topographic structure built by accumulated deposits of volcanic eruptions; a mountain built by eruptions.

erosion. The physical removal of weathered rock and sediment by water, wind, glaciation, or gravity.

exfoliate. To flake apart in thin sheets.

fan. A laterally and vertically spreading deposit of sediment left by a stream or landslide at the mouth of a canyon.

fault. A fracture along which two pieces of crust move relative to each other. Movement on faults generates earthquakes.

fault zone. A region of the crust with parallel or branching faults.

feldspar. The most common group of felsic rock-forming minerals. Feldspars are light-colored silicates with aluminum plus varying amounts of one or more of calcium, sodium, or potassium. Cleavage planes intersect at 90° angles.

felsic. Descriptive term for (typically) light-colored igneous rocks such as dacite, rhyolite, and granite that are enriched in feldspar and other silica-rich minerals. The term is also applied to the high-silica minerals in these rocks. *See also* mafic.

fissure. An elongated volcano or closely spaced alignment of volcanoes where a dike reaches the surface. Eruptions usually occur at more than one location simultaneously.

flow field. An accumulation of multiple lava flows of similar composition that erupted from the same vent, though not necessarily close together in time.

fluid. A substance that is able to move or change shape without separating into fragments. Solid rock can behave as a fluid over very large time intervals if temperature or pressure is sufficient.

foliation. Layering characterized by the flattening or elongation of minerals or clasts due to pressure during metamorphism.

formation. A formally named body of rock strata that are related by depositional history. A formation is distinctive from other strata in a region and is extensive enough to be mapped, both on the surface and in the subsurface.

glacial drift. *See also* drift.

glacial erratic. A clast (typically a boulder) transported and deposited by ice that is a different rock type than the rock on which it rests.

glacial maximum. The time during a period of glacial advance when glaciers are at their greatest extent. *See also* glaciation.

glacial outwash. *See* outwash.

glacial till. *See* till.

glaciation. The formation and movement of glaciers. A climatic period during which alpine or continental glaciers formed, reached a glacial maximum, and receded.

glaciomarine drift. Sediment that falls to the seafloor from floating ice.

glass. Rapidly chilled lava or magma that has no crystalline structure.

gneiss. A foliated metamorphic rock produced under great pressure and heat, often characterized by alternating bands of flattened or stretched light and dark minerals.

GPS (global positioning system). A satellite-based navigation system made up of a network of satellites placed into orbit by the US Department of Defense.

grain. A rock or mineral particle.

granite. A light-colored, medium- to coarse-grained, silica-rich (felsic) intrusive igneous rock with lesser amounts of dark, mafic minerals, such as biotite or hornblende. The plutonic equivalent of rhyolite. Also used imprecisely as a generic term for coarse-grained igneous rocks.

granitic. Composed of granite. More generally applied to plutonic rocks that are similar to granite, such as granodiorite, quartz diorite, tonalite, and quartz monzonite.

granodiorite. A light-colored, medium- to coarse-grained plutonic rock. Its silica content is intermediate between granite and diorite.

gravel. A deposit consisting of rounded, unconsolidated rock fragments between 0.2 and 3 inches (5 and 76 mm) across. Also refers to individual clasts of that size.

greenschist. A foliated metamorphic rock formed at low pressures and temperatures. The color is due to green minerals, usually chlorite, epidote, or actinolite.

greenstone. A nonfoliated metamorphic rock formed at low pressures and temperatures. The protolith is mafic volcanic rock.

groundmass. Glass or microcrystals surrounding phenocrysts in lava.

group. A grouping of several different formations of similar age.

hornblende. A dark, rock-forming mafic mineral common in felsic igneous and metamorphic rocks. A member of the amphibole group of minerals, it is recognized by its splintery appearance and alternating 60° and 120° angles between its four crystal faces.

hummock. A mound of unconsolidated rock fragments above a generally level surface.

hypersthene. A beer-bottle brown variety of orthopyroxene commonly found in Cascade Range andesite and dacite.

hypothesis. A proposed explanation of a natural phenomenon when only limited evidence is available. The hypothesis is the starting point for further investigation. It is not assumed to be true. Additional evidence may lead to a theory.

ice age. A prolonged period of cold climate distinguished by the periodic advance of glaciers.

ice sheet. An extensive glacier with a low-gradient surface. Often formed by the convergence of smaller glaciers.

igneous. A large class of rocks formed by the cooling of magma.

ignimbrite. A pumice-rich volcanic deposit that may or may not be welded. *See also* welded tuff.

interglacial. An interval of time between glacial periods.

intrusion. A body of intrusive igneous rock of any size that invades older rocks. Examples are plutons and dikes.

ion. An atom or molecule carrying a positive or negative charge.

isotope. One of two or more species of the same chemical element that have the same number of protons but a different number of neutrons. *See also* radioisotope.

joint. A rock fracture across which no relative movement occurs.

katabatic. A wind that carries high-density cold air down a slope under the force of gravity.

knickpoint. An interruption or break in the general gradient of a stream, particularly a point of abrupt change. Waterfalls or very steep rapids are found at knickpoints.

lag deposit. The coarse grains left behind after finer-grained sediment is carried away by water or wind.

lahar. A debris flow resulting from an eruption or originating in volcanic deposits.

lapilli. Tephra clasts between 0.08 and 2.5 inches (2 and 64 mm) across.

lava. Magma extruded onto the surface, or its solidified product.

lava flow. An outpouring of magma from a volcano, or the resulting cooled volcanic structure.

limestone. A sedimentary rock composed of calcium carbonate, often in the form of marine shells.

lithify. The conversion of sediment to solid rock by compaction, chemical cementation, or crystallization.

lithosphere. Earth's rigid, solid shell. It consists of the crust and the outermost portion of the mantle.

loam. A rich, permeable soil composed of sand, silt, clay, and organic matter.

longshore drift. The transport of sediment parallel to a coast by ocean currents.

mafic. Descriptive term for (typically) dark-colored igneous rocks, such as gabbro and basalt, that are enriched in magnesium or iron-bearing minerals, including olivine and clinopyroxene. Also the low-silica minerals that define these rocks. *See also* felsic.

magma. A mixture of liquid rock, solid crystals, and dissolved gases that cools beneath the surface to form plutonic rocks or erupts at the surface to form volcanic rocks.

mantle. A layer in Earth's interior between the crust and the outer core. The mantle is hot but solid rock that deforms like plastic over long periods of time.

marble. Metamorphosed limestone.

metamorphism. The mineralogical, chemical, and structural changes that occur in rock in a solid (not liquid) state when it is subjected to heat or pressure.

microcrystalline. A rock texture characterized by microscopic crystals.

migmatite. A mix of metamorphic rock, usually gneiss, and intrusive igneous rocks. Usually has a swirled appearance.

mineral. A naturally occurring, crystalline, inorganic solid with characteristic chemical composition, atomic structure, and physical properties, such as hardness, crystal shape, density, and color.

moraine. A ridge-shaped accumulation of glacial till deposited at the margins of a glacier.

mud. A mix of sand-, silt-, or clay-sized grains.

normal fault. An inclined fault along which the upper rock body moves down relative to the lower rock body. Movement is the result of crustal extension, or spreading. *See also* reverse fault, strike-slip fault, and thrust fault.

olivine. A green mineral common in mafic igneous rocks. Olivine is among the first minerals to crystallize in magma and is a principal component of the upper mantle.

orthopyroxene. A group of usually light-colored silicate minerals in the pyroxene family, differentiated from clinopyroxene by details in crystal arrangement.

outwash. Alluvium deposited by meltwater flowing directly from glaciers.

pahoehoe. Basalt lava that takes the form of smooth sheets or swirled ridges. *See also* aa.

Pangaea. A supercontinent that existed from about 300 to 200 million years ago. It included most of Earth's continents, which fragmented to form the present configuration of continents.

Panthalassa Ocean. The global ocean that surrounded Pangaea.

pebble. A rock fragment, typically at least partly rounded by abrasion, with a diameter between about 0.2 and 2.5 inches (4 and 64 mm).

peridotite. An igneous rock that crystallizes from magma consisting mostly of the mineral olivine.

permafrost. A zone in the soil that remains frozen year-round.

phenocryst. A mineral crystal in igneous rocks that is visible to the naked eye. *See also* microcrystalline.

phyllite. A fine-grained metamorphic rock characterized by wavy or kinky foliation. The protolith is shale or mudstone.

pillow lava. A globular form taken by lava that flows underwater.

plagioclase. A type of feldspar that is usually transparent and colorless, taking the form of tabular, square-sided prisms. Nearly ubiquitous in Cascade Range igneous rocks.

planar. Flat or sheetlike.

plate. A portion of the lithosphere that slides independently on top of the asthenosphere. Plates may diverge, converge, or move laterally relative to each other.

plate tectonics. A theory that posits Earth's lithosphere is broken into plates that move and interact with each other.

pluton. An igneous intrusion that cools within the crust and has a relatively uniform composition. Plutons are the components of batholiths.

plutonic. Igneous rocks that cool within the crust. Crystals are typically 0.8 inch (2 mm) or larger and visible to the naked eye.

porphyritic. A texture of volcanic rocks characterized by minerals visible to the naked eye surrounded by a matrix of glass or microcrystals. *See also* microcrystalline, plutonic, and texture.

protolith. The unmetamorphosed "parent" rock of a metamorphic rock.

provenance. The source area for an erratic or sediment, determined by distinctive chemistry, fossils, composition, or other features.

Puget Lobe. A lobe of the Cordilleran Ice Sheet that extended southward into the Puget Lowland.

Puget Lowland. The low-lying region situated between the Cascade Range to the east, the Olympic Mountains and Willapa Hills to the west, the San Juan Islands to the north, and the Columbia River to the south. Much of the lowland is filled by the Salish Sea.

pumice. Felsic tephra that is at least 50 percent vesicles by volume. *See also* scoria.

pyroclastic flow. An incandescent flow of gas, ash, and volcanic rock that flows downslope during a volcanic eruption. *See also* ignimbrite.

pyroclastic surge. High-velocity superheated air and some clastic material that spreads beyond a pyroclastic flow or lateral eruption blast. Analogous to the blast of air and dust that extends beyond a landslide.

pyroxene. A group of mafic, rock-forming silicate minerals common in Cascade Volcanic Arc basalt and andesite. Crystals may be recognized by right-angle cleavage. *See also* clinopyroxene and orthopyroxene.

quartz. A common, clear, white, or grayish rock-forming mineral composed solely of silica.

quartz diorite. A black-and-white-speckled, medium- to coarse-grained intrusive igneous rock similar to diorite but containing quartz in addition to the minerals characterizing diorite.

radiocarbon. The carbon isotope carbon-14.

radioisotope. A radioactive isotope of an element that decays to a stable isotope. The abundance ratio of certain radioisotopes can be used to determine the age of some minerals.

reverse fault. A fault, along which the upper rock body moves upward relative to the lower rock body. It is steeper than a thrust fault and the opposite of a normal fault.

rhyolite. A rock characterized by phenocrysts of quartz and feldspar in a fine-grained groundmass. The most silica-rich volcanic rock. The volcanic equivalent of granite.

rift. A long fracture in the lithosphere, usually on continents, that develops where tectonic plates are moving away from each other. Volcanism is common in rift zones. *See also* spreading center.

rock. A naturally occurring mass of minerals or rock clasts that results from the cementation of individual grains or the cooling of magma. *See also* urbanite.

Salish Sea. The official geographic term referring to the inland marine waters of Washington and British Columbia, from the north end of the Strait of Georgia and Desolation Sound to the south end of Puget Sound, as well as the Strait of Juan de Fuca.

sand. Grains of sediment between 0.0025 and 0.08 inch (0.06 and 2 mm) across.

scarp. A cliff face formed by erosion, such as that related to wave action, landslides, or faulting. Fault scarps tend to be ephemeral due to erosion and bioturbation.

scoria. A mafic volcanic rock that is rich in (usually) large and prominent vesicles. Scoria is erupted at cinder cones. *See also* pumice.

seamount. An undersea volcano that does not breach the surface.

sea stack. A small island near the shore that results from wave erosion of coastal headlands and bluffs.

sediment. Unconsolidated material transported by gravity, water, ice, or wind or chemically secreted by organisms. Sediment is deposited in layers and may eventually be lithified. Some geologists include tephra, as well.

sedimentation. The process of the deposition of sediment.

shale. A fine-grained sedimentary rock that results from the lithification of mud or silt.

silica. A molecule containing atoms of silicon and oxygen. The most common molecule in Earth's rocks. Silica molecules bond together to form the basic crystalline structure of silicate rocks.

silicate. A mineral that is based on the molecular structure of silica.

sill. A tabular igneous intrusion that follows bedding planes in the rocks it invades. *See also* dike.

silt. A sedimentary particle between 0.00016 and 0.0025 inch (0.004 and 0.06 mm) across that is finer than sand and coarser than clay. Also fine sediment suspended in water.

slate. A foliated metamorphic rock formed at low pressure and temperature. It is characterized by foliation along which the rock splits into thin sheets. The protolith is shale.

sorting. Separation of minerals or sedimentary clasts by density or size.

spreading center. A place in the lithosphere, usually submarine, where two tectonic plates move away from each other. Characterized by eruptions of magma rising from the mantle.

strata. Layers of sedimentary rock.

stratigraphy. The scientific study of rock strata, including interpretation of origins, ages, and relationships. Also the arrangement of strata in a geologic exposure.

stratovolcano. A volcanic edifice built of successive lava flows, sometimes interspersed with pyroclastic flows and tephra. Generally associated with viscous lava that piles up close to a vent rather than that which flows a great distance.

striation. One of multiple scratches or grooves eroded into rock by the rock debris carried at the base of a glacier.

strike-slip fault. A fault in which the rock bodies on either side move laterally relative to each other rather than vertically.

structure. The physical arrangement of rock strata, faults, intrusions, crystals, and other geologic features.

subduction. The process in which an upper plate (the subducting plate) overrides a lower plate (the subducted plate) in a subduction zone.

subduction complex. Oceanic sediment and rock strata, often unrelated and of different ages, scraped off a subducted oceanic plate and brought together on the overriding plate.

subduction zone. The zone where one tectonic plate overrides another at a convergent plate boundary. Subduction zones are characterized by thrust faulting and volcanic arcs on the upper plate.

subsidence. Downward settling or sinking of some portion of the Earth's surface.

superterrane. A grouping of terranes sharing a distinctive tectonic history.

tectonics. Pertaining to the large-scale processes that deform Earth's crust. *See also* plate tectonics.

tephra. Rock or lava erupted into the atmosphere during an eruption. Tephra is subdivided by particle size into ash, lapilli, and bombs.

terminus. The end of a glacier farthest from the glacier's source area.

terrane. A crustal block or fragment, bounded by faults, that has a different geologic history from adjacent rocks. It is inferred that a terrane was moved by plate tectonics from a former position.

texture. The physical characteristics of a rock, including the size and arrangement of grains and crystals; the size, shape, and number of vesicles; and layering.

theory. An explanation or model based on observation, experimentation, and reasoning that has been tested and confirmed as a general principle. A theory helps explain and predict natural phenomena but may be replaced by later theories.

thrust fault. A fault resulting from compression along which the upper rock body moves over the lower rock body. The opposite of a normal fault. Subduction zones are characterized by thrust faults. Thrust faults are generally low angle—45° or less—relative to steeper reverse faults.

till. Unsorted glacial drift deposited directly by melting ice.

tonalite. A light-colored, medium- to coarse-grained intrusive rock that is compositionally similar to diorite but with less quartz. The plutonic equivalent of andesite.

transform fault. A strike-slip fault that is broken into segments and offsets a divergent plate boundary.

travertine. A calcareous rock without structure precipitated from solution at springs and in limestone caves.

tsunami. A wave generated by subsidence or uplift of the seafloor during an earthquake.

tuff. A volcanic rock that is principally lithified ash or a mixture of ash and larger rock clasts. Tuff breccia is tuff with a high proportion of large clasts.

turbidite. A sedimentary rock that is the lithified deposit of a turbidity current.

turbidity current. The flow of suspended sediment along or just above the seafloor.

unconformity. A gap in the geologic record due to erosion or the lack of sediment deposition and recognized by rocks of demonstrably different ages or types lying in direct contact with each other.

uplift. Upward movement of the lithosphere either by rebound after glaciers have melted, the addition of magma, or thrust faulting.

urbanite. A tongue-in-cheek term for a man-made rocky substance, such as concrete, asphalt, or terra-cotta.

vent. An opening at Earth's surface from which magma erupts.

vesicle. A gas bubble preserved in cooled volcanic rock. *See also* scoria and pumice.

viscosity. The measure of a substance's resistance to flow. Substances with higher viscosity, such as honey or rhyolite, move more slowly and spread out less than lower-viscosity substances, such as water or basalt.

volcanic arc. A linear arrangement of volcanoes above a subduction zone.

volcanic field. A mostly contiguous collection of volcanic rocks, often erupted from different volcanic vents over a span of time.

volcano. A generic term for the place where magma reaches Earth's surface. May or not grow into a mountain. *See also* caldera, cinder cone, stratovolcano, and vent.

weathering. The disintegration and decomposition of rock by chemical or physical processes at or very near Earth's surface.

welded tuff. Ignimbrite that was hot enough at emplacement that pumice and ash clasts were compressed, lithified, and deformed by the overlying mass of the deposit.

zircon. A very hard, silicate mineral found in many rocks and useful for age dating.

Sources of More Information

General Reading

Bates, R. L., and J. A. Jackson, eds. 1984. *Dictionary of Geological Terms,* 3rd edition. American Geological Institute. New York: Anchor Books.

Carson, B., and S. Babcock. 2009. *Hiking Guide to Washington Geology.* Sandpoint, ID: Keokee Publishing.

Harris, S. L. 2005. *Fire Mountains of the West: The Cascade and Mono Lake Volcanoes,* 3rd edition. Missoula, MT: Mountain Press Publishing Co.

Hildreth, W. 2007. *Quaternary Magmatism in the Cascades — Geologic Perspectives.* US Geological Survey Professional Paper 1744. pubs.usgs .gov/pp/pp1744/.

Kiver, E. In press. *Washington Rocks!* Missoula, MT: Mountain Press Publishing Co.

Lynch, D. R., and B. Lynch. 2012. *Rocks and Minerals of Washington and Oregon: A Field Guide to the Evergreen and Beaver States.* Cambridge, MN: Adventure Publications.

Mathews, B., and J. Monger. 2005. *Roadside Geology of Southern British Columbia.* Missoula, MT: Mountain Press Publishing Co.

Romaine, G. 2012. *Rocks, Gems, and Minerals: A Falcon Field Guide.* Guilford, CT: Globe Pequot Press.

Romaine, G. 2007. *Gem Trails of Washington.* Upland, CA: Gem Guides Book Co.

Geologic Maps

Geologic maps for all of Washington at a meaningful scale are not available. The best online source for topographic-based geologic maps at the 7.5-minute scale (1:24,000) is www.dnr.wa.gov/ Publications/ger_24k_mapping_status.pdf.

Maps at the 1:100,000 scale, covering the entire state, are available at www.dnr.wa.gov/Publications/ger_geologic_maps_wa.pdf

Washington geologic maps at other scales are listed in a catalog at www .dnr.wa.gov/publications/ger_publications_list.pdf.

FIELD GUIDES

A list of geologic field guides for the amateur and professional is available at: www.dnr.wa.gov/Publications/ger_geologic_field_trip_guides_list.pdf.

INTRODUCTION

Beck, M. E., and L. Noson. 1972. "Anomalous Paleolatitudes in Cretaceous Granitic Rocks." *Nature Physical Science* 235: 11–13.

Brown, E. H., and J. D. Dragovich. 2003. *Tectonic Elements and Evolution of Northwest Washington*. Washington Division of Geology and Earth Resources Geologic Map GM-52, 1 sheet, scale 1:625,000, with 12 pages of text. www.dnr.wa.gov/Publications/ger_gm52_tectonic_evolution_nw_washington.zip.

Housen, B. A., and M. E. Beck, Jr. 1999. "Testing Terrane Transport: An Inclusive Approach to the Baja B. C. Controversy." *Geology* 27: 1143–46.

Kious, W. J., and R. I. Tilling. 1996. *This Dynamic Earth: The Story of Plate Tectonics*. US Government Printing Office. Available online with periodic updates. pubs.usgs.gov/gip/dynamic/dynamic.html.

Molenaar, D. 1988. *The Spokane Aquifer, Washington: Its Geologic Origin and Water-Bearing and Water-Quality Characteristics*. US Geological Survey Water Supply Paper 2265. pubs.er.usgs.gov/publication/wsp2265.

Scotese, C. R. "PALEOMAP Project." Scotese.com.

Tabor, R. W., and R. A. Haugerud. 1999. *Geology of the North Cascades: A Mountain Mosaic*. Seattle, WA: The Mountaineers.

Umhoefer, P. J., and R. C. Blakey. 2006. "Moderate (1,600 km) Northward Translation of British Columbia from Southern California: An Attempt at Reconciliation of Paleomagnetism and Geology." In *Paleogeography of the North American Cordillera: Evidence for and Against Large-Scale Displacements*, Geological Association of Canada Special Paper 46, edited by J. W. Haggart, R. J. Enkin, and J. W. H. Monger, 305–27.

Wells, R. E., C. S. Weaver, and R. J. Blakely. 1998. "Fore-Arc Migration in Cascadia and Its Neotectonic Significance." *Geology* 26: 759–62. courses.washington.edu/ess403/CascadiaResources/WellsGSA98.pdf.

Williams, H. 2002. *The Restless Northwest: A Geological Story*. Pullman, WA: Washington State University Press.

Wilson, J. W., and R. Clowes. 2009. *Ghost Mountains and Vanished Oceans: North America from Birth to Middle Age*. Toronto: Key Porter Books.

1. The Bridge of the Gods and the Bonneville Dam

Moulton, G. E., ed. 2002. *The Definitive Journals of Lewis and Clark*, vols. 2–8. Lincoln: University of Nebraska Press.

Norman, D. K., and J. M. Roloff. 2004. *A Self-Guided Tour of the Geology of the Columbia River Gorge – Portland Airport to Skamania Lodge, Stevenson, Washington.* Olympia: Washington Division of Geology and Earth Resources Open File Report 2004–7. www.dnr.wa.gov /Publications/ger_ofr2004-7_geol_tour_columbia_river_gorge.pdf.

O'Connor, J. E. 2004. "The Evolving Landscape of the Columbia River Gorge: Lewis and Clark and Cataclysms on the Columbia." *Oregon Historical Quarterly* 105: 390–421. ohs.org/research/quarterly /images/OHQ1053_OConnor.pdf.

O'Connor, J. E., and S. F. Burns. 2009. "Cataclysms and Controversy – Aspects of the Geomorphology of the Columbia River Gorge." In *Volcanoes to Vineyards: Geologic Field Trips through the Dynamic Landscape of the Pacific Northwest*, edited by J. E. O'Connor, R. J. Dorsey, and I. Madin, 237–51. Boulder, CO: Geological Society of America.

University of Nebraska Press/University of Nebraska-Lincoln Libraries-Electronic Text Center. *The Journals of the Lewis and Clark Expedition.* lewisandclarkjournals.unl.edu.

2. Beacon Rock and the Missoula Floods

Evarts, R. C., R. M. Conrey, R. J. Fleck, and J. T. Hagstrum. 2009. "The Boring Volcanic Field of the Portland-Vancouver Area, Oregon and Washington: Tectonically Anomalous Forearc Volcanism in an Urban Setting." In *Volcanoes to Vineyards: Geologic Field Trips through the Dynamic Landscape of the Pacific Northwest*, edited by J. E. O'Connor, R. J. Dorsey, and I. Madin, 253–70. Boulder, CO: Geological Society of America.

Hay, K. G. 2004. *The Lewis and Clark Columbia River Water Trail: A Guide for Paddlers, Hikers, and Other Explorers.* Portland, OR: Timber Press.

University of Nebraska Press/University of Nebraska-Lincoln Libraries-Electronic Text Center. *The Journals of the Lewis and Clark Expedition.* lewisandclarkjournals.unl.edu.

3. Cape Disappointment

Allan, J. C., R. C. Witter, P. Ruggiero, and A. D. Hawkes. 2009. "Coastal Geomorphology, Hazards, and Management Issues along the Pacific Northwest Coast of Oregon and Washington." In *Volcanoes to Vineyards: Geologic Field Trips through the Dynamic Landscape of the Pacific Northwest*, edited by J. E. O'Connor, R. J. Dorsey, and I. Madin, 495–519. Boulder, CO: Geological Society of America.

Gelfenbaum, G., and G. M. Kaminsky. 2010. "Large-Scale Coastal Change in the Columbia River Littoral Cell: An Overview." *Marine Geology* 273: 1–10.

Hayes, D. 1999. *Historical Atlas of the Pacific Northwest: Maps of Exploration and Discovery.* Seattle, WA: Sasquatch Books.

Kaminsky, G. M., P. Ruggiero, M. C. Buijsman, D. S. McCandless, and G. Gelfenbaum. 2010. "Historical Evolution of the Columbia River Littoral Cell." *Marine Geology* 273: 96–126.

Twitchell, D. C., and V. A. Cross. 2001. *Holocene Evolution of the Southern Washington and Northern Oregon Shelf and Coast: Geologic Discussion and GIS Data Release.* US Geological Survey Open-File Report 01-076. pubs.usgs.gov/of/2001/of01-076/index.htm.

4. The Sunken Forest of Willapa Bay

Atwater, B. F., M. R. Satoko, K. Satake, T. Yoshinobu, U. Kazue, and D. Yamaguchi. 2005. *The Orphan Tsunami of 1700: Japanese Clues to a Parent Earthquake in North America.* US Geological Survey Professional Paper 1707. pubs.usgs.gov/pp/pp1707/.

Atwater, B. F., M. Stuiver, and D. K. Yamaguchi. 1991. "Radiocarbon Test of Earthquake Magnitude at the Cascadia Subduction Zone." *Nature* 353: 156–58.

Cascadia Region Earthquake Workgroup. 2013. *Cascadia Subduction Zone Earthquakes: A Magnitude 9.0 Earthquake Scenario.* www.dnr.wa.gov/publications/ger_ic116_csz_scenario_update.pdf.

Doughton, S. 2013. *Full Rip 9.0.* Seattle: Sasquatch Books.

Hyndman, R. D., G. C. Rogers, H. Dragert, K. Wang, J. J. Clague, J. Adams, and P. T. Bobrowsky. 1996. "Giant Earthquakes beneath Canada's West Coast." *Geoscience Canada* 23: 63–72.

Kroeber, A. L. 1976. *Yurok Myths.* Berkeley: University of California Press.

Ludwin, R. S., R. Dennis, D. Carver, A. D. McMillan, R. Losey, J. Clague, C. Jonientz-Trisler, J. Bowechop, J. Wray, and K. James. 2005. "Dating the 1700 Cascadia Earthquake: Great Coastal Earthquakes in Native Stories." *Seismological Research Letters* 76: 140–48.

NOAA Center for Tsunami Research, Pacific Marine Environmental Laboratory. "Animations." nctr.pmel.noaa.gov/animate.html.

Thompson, J. 2011. *Cascadia's Fault: The Earthquake and Tsunami That Could Devastate North America.* Berkeley, CA: Counterpoint Press.

5. Columbia River Basalt in the Chehalis River

Benton, M. J. 2003. *When Life Nearly Died: The Greatest Mass Extinction of All Time.* London: Thames and Hudson.

Hooper, P. R. 2000. "Flood Basalts Provinces." In *Encyclopedia of Volcanoes*, edited by H. Sigurdsson, B. Houghton, S. R. McNutt, H. Rymer, and J. Stix, 345–59. San Diego, CA: Academic Press (Elsevier).

Reidel, S. P. 2005. "A Lava Flow without a Source: The Cohassett Flow and Its Compositional Components, Sentinel Bluffs Member, Columbia River Basalt Group." *Journal of Geology* 113: 1–21.

Self, S., T. Thordarson, L. Keszthelyi, G. P. L. Walker, K. Hon, M. T. Murphy, P. Long, and S. Finnemore. 1996. "A New Model for the Emplacement of Columbia River Basalts as Large, Inflated Pahoehoe Lava Flow Fields." *Geophysical Research Letters* 23: 2689–92.

Shaw, H. R., and D. A. Swanson. 1970. "Eruption and Flow Rates of Flood Basalts." In *Proceedings of the Second Columbia River Basalt Symposium*, edited by E. H. Gilmour and D. Straddling, 271–99. Cheney: Eastern Washington State College Press.

Tolan, T. L., B. S. Martin, S. P. Reidel, J. L. Anderson, K. A. Lindsey, and W. Burt. 2009. "An Introduction to the Stratigraphy, Structural Geology, and Hydrogeology of the Columbia River Flood-Basalt Province: A Primer for the GSA Columbia River Basalt Group Field Trips." In *Volcanoes to Vineyards: Geologic Field Trips through the Dynamic Landscape of the Pacific Northwest*, edited by J. E. O'Connor, R. J. Dorsey, and I. Madin, 599–643. Boulder, CO: Geological Society of America.

6. THE MIMA MOUNDS

Berg, A. W., 1990. "Formation of Mima Mounds: A Seismic Hypothesis." *Geology* 18: 281–84.

Cox, G. W. 1984. "The Distribution and Origin of Mima Mound Grasslands in San Diego County, California." *Ecology* 65: 1397–405.

Cox, G. W., and A. W. Berg. 1990. "Comment and Reply on 'Formation of Mima Mounds: A Seismic Hypothesis.'" *Geology* 18: 1259–61.

Gabet, E .J., J. T. Perron, and D. L. Johnson. 2014. "Biotic Origin for Mima Mounds Supported by Numerical Modeling." *Geomorphology* 206: 58–66.

Goldstein, B. S., P. T. Pringle, B. Parker, and Z. O. Futornick. 2010. "Tracking the Late-Glacial Outburst Flood from Glacial Lake Carbon, Washington State, USA." In *Northwest Scientific Association 82nd Annual Meeting*, program and abstracts, 57. www.centralia .edu/academics/earthscience/nwsa/NWSA_CPOP_2010_ Program.pdf. Link to poster: www.centralia.edu/academics /earthscience/pringle/pubs/Tanwax_NWSA_poster2010.pdf

Jackson, H. E. 1956. "The Mystery of the Mima Mounds." *Natural History* 65: 136–39.

Logan, R. L., and T. J. Walsh. 2009. "Mima Mounds Formation and Their Implications for Climate Change." In *Northwest Scientific Association 81st Annual Meeting*, 38–39.

Saucier, R. T., and A. W. Berg. 1991. "Comment and Reply on 'Formation of Mima Mounds: A Seismic Hypothesis.'" *Geology* 19: 284–85.

Washburn, A. L. 1988. *Mima Mounds: An Evaluation of Proposed Origins with Special Reference to the Puget Lowland*. Washington Division of Geology and Earth Resources Report of Investigations 29. www.dnr .wa.gov/Publications/amp_na_mima_mounds_1988.pdf.

7. MOUNT ST. HELENS NATIONAL VOLCANIC MONUMENT

Greeley, R., and J. H. Hyde. 1972. "Lava Tubes of the Cave Basalt, Mount St. Helens, Washington." *Geological Society of America Bulletin* 83: 2397–418.

Halliday, W. R. 1983. *Ape Cave and Mount St. Helens Apes*. Vancouver, WA: ABC Publishing.

Hyde, J. H., and R. Greeley. 1973. "Geological Field Trip Guide, Mount Saint Helens Lava Tubes." In *Geologic Fields Trips in Northern Oregon and Southern Washington*, Oregon Department of Geology and Mineral Industries Bulletin 77, edited by J. D. Beaulieu, 183–206.

Larson, C. V. 1993. *An Illustrated Glossary of Lava Tube Features*. Western Speleological Survey Bulletin 87. Vancouver, WA: ABC Publishing.

Orr, T. R. 2011. "Lava Tube Shatter Rings and Their Correlation with Lava Flux Increases at Kilauea, Hawaii." *Bulletin of Volcanology* 73: 335–46.

Pringle, P. T. 1993. *Roadside Geology of the Mount St. Helens National Volcanic Monument and Vicinity*. Washington Department of Natural Resources, Division of Geology and Earth Resources Information Circular 88. Available online in two parts: www.dnr.wa.gov /Publications/ger_ic88_mount_st_helens_pt1.pdf, and www.dnr .wa.gov/Publications/ger_ic88_mount_st_helens_pt2.pdf.

Thornber, C. R. 2001. "Olivine-Liquid Relations of Lava Erupted by Kilauea Volcano from 1994 to 1998: Implications for Shallow Magmatic Processes Associated with the Ongoing East-Rift-Zone Eruption," *Canadian Mineralogist* 39: 239–66.

Williams, D. A., S. D. Kadel, R. Greeley, C. M. Lesher, and M. A. Clynne. 2004. "Erosion by Flowing Lava: Geochemical Evidence in the Cave Basalt, Mount St. Helens, Washington." *Bulletin of Volcanology* 66: 168–81.

8. JOHNSTON RIDGE

Hickson, C. J. 2005. *Mount St. Helens: Surviving the Stone Wind.* Vancouver, BC: Tricouni Press.

Lipman, P. W., and D. R. Mullineaux, eds. 1981. *The 1980 Eruptions of Mount St. Helens, Washington.* US Geological Survey Professional Paper 1250.

Pringle, P. T. 1993. *Roadside Geology of the Mount St. Helens National Volcanic Monument and Vicinity.* Washington Department of Natural Resources, Division of Geology and Earth Resources Information Circular 88. Available online in two parts: www.dnr.wa.gov /Publications/ger_ic88_mount_st_helens_pt1.pdf, and www.dnr .wa.gov/Publications/ger_ic88_mount_st_helens_pt2.pdf.

Sherrod, D. R., W. E. Scott, and P. H. Stauffer, eds. 2008. *A Volcano Rekindled: The Renewed Eruption of Mount St. Helens, 2004–2006.* US Geological Survey Professional Paper 1750.

9. SUNRISE, MOUNT RAINIER NATIONAL PARK

There are more hiking guidebooks than you can shake several sticks at for the trails in Mount Rainier National Park. The only one I know of with reliable geology is Carson and Babcock, 2009. Pat Pringle's excellent roadside guide has mile-by-mile descriptions of roads in and around the park and is an essential companion to fans of Mount Rainier geology.

Carson, B., and S. Babcock. 2009. *Hiking Guide to Washington Geology.* Sandpoint, ID: Keokee Publishing.

Driedger, C. L., and W. E. Scott. 2008. *Mount Rainier – Living Safely with a Volcano in Your Backyard.* US Geological Survey Fact Sheet 2008–3062. pubs.usgs.gov/fs/2008/3062/fs2008-3062.pdf.

Pringle, P. T. 2008. *Roadside Geology of Mount Rainier National Park and Vicinity.* Washington Division of Geology and Earth Resources Information Circular 107. ftp://ww4.dnr.wa.gov/geology/pubs /ic107/parts/ic107_mt_rainier_guide_intro.pdf.

Scott, K. M., J. W. Vallance, and P. T. Pringle. 1995. *Sedimentology, Behavior, and Hazards of Debris Flows at Mount Rainier, Washington.* US Geological Survey Professional Paper 1547. pubs.er.usgs.gov /publication/pp1547.

Sisson, T. W., and J. W. Vallance. 2009. "Frequent Eruptions of Mount Rainier over the Last 2,600 Years." *Bulletin of Volcanology* 71: 595–618. link.springer.com/article/10.1007%2Fs00445-008-0245-7?LI=true.

Stockstill, K. R., T. A. Vogel, and T. W. Sisson. 2002. "Origin and Emplacement of the Andesite of Burroughs Mountain, a Zoned, Large-Volume Lava Flow at Mount Rainier, Washington, USA." *Journal of Volcanology and Geothermal Research* 119: 275–96.

10. Nisqually Glacier

Carson, B., and S. Babcock. 2009. *Hiking Guide to Washington Geology.* Sandpoint, ID: Keokee Publishing.

Driedger, C., T. Sisson, and J. Vallance. 2014. *Journey Back in Time: A Mount Rainier Geological Field Trip Guide for Teachers,* Nisqually Entrance to Paradise. US Geological Survey GIP 19. pubs.usgs.gov /gip/19/downloads/Appendixes/III_Journey%20Back%20in%20 Time-Field%20Guide.pdf.

Harrison, A. E. 1956. "Fluctuations of the Nisqually Glacier, Mt. Rainier, Washington, since 1750." *Journal of Glaciology* 2: 675–83.

Huybers, P. J., and C. Langmuir. 2009. "Feedback between Deglaciation, Volcanism, and Atmospheric CO_2." *Earth and Planetary Science Letters* 286:479–91. dash.harvard.edu/bitstream/handle/1/3659701 /Huybers_FeedbackDeglaciation.pdf?sequence=1.

National Park Service. North Cascades National Park. "Glacier Monitoring." ww.nps.gov/noca/naturescience/glacial-mass-balance1.htm.

Nylen, T. H. 2004. "Spatial and Temporal Variations of Glaciers on Mt. Rainier between 1913 and 1994." Master's thesis, Department of Geology, Portland State University, OR.

Porter, S. C. "Recent Glacier and Climate Variations in the Pacific Northwest." Quaternary Research Center, University of Washington. faculty.washington.edu/scporter/Rainierglaciers.html.

Pringle, P. 2008. *Roadside Geology of Mount Rainier National Park and Vicinity.* Washington Division of Geology and Earth Resources Information Circular 107. ftp://ww4.dnr.wa.gov/geology/pubs /ic107/parts/ic107_mt_rainier_guide_intro.pdf.

Sanford, J. E. 2011. "Glacial Changes between 1985–2009 and Implications for Volcanic Hazards at Mt. Rainier, Washington." Master's thesis, Oklahoma State University, Stillwater, OK. dc.library.okstate. edu/utils/getfile/collection/theses/id/4109/filename/4110.pdf.

Scott, K. M., J. W. Vallance, and P. T. Pringle. 1995. *Sedimentology, Behavior, and Hazards of Debris Flows at Mount Rainier, Washington.* US Geological Survey Professional Paper 1547. pubs.er.usgs.gov /publication/pp1547.

Veatch, F. M. 1969. *Analysis of a 24-Year Photographic Record of Nisqually Glacier, Mount Rainier National Park, Washington.* US Geological Survey Professional Paper 631. pubs.er.usgs.gov/publ ication/pp631.

Walder, J. S., and C. L. Driedger. 1994. *Geomorphic Change Caused by Outburst Floods and Debris Flows at Mount Rainier, Washington, with Emphasis on Tahoma Creek Valley.* US Geological Survey Water-Resources Investigations Report 93–4093. pubs.er.usgs.gov /publication/wri934093.

11. Whidbey Island

Booth, D. B., K. G. Troost, J. J. Clague, and R. B. Waitt. 2004. "The Cordilleran Ice Sheet." In *The Quaternary Period in the United States*, edited by A. R. Gillespie, S. C. Porter, and B. F. Atwater, 17–44. Amsterdam, The Netherlands: Elsevier.

Easterbrook, D. J. 1994. "Stratigraphy and Chronology of Early to Late Pleistocene Glacial and Interglacial Sediments in the Puget Lowland, Washington." In *Geologic Field Trips in the Pacific Northwest*, vol. 1, edited D. A. Swanson and R. A. Haugerud, 1–38. Boulder, CO: Geological Society of America.

Easterbrook, D. J., D. R. Crandell, and E. B. Leopold. 1967. "Pre-Olympia Pleistocene Stratigraphy and Chronology in the Central Puget Lowland, Washington." *Geological Society of America Bulletin* 78: 13–20.

Hagstrum, J. T., D. B. Booth, K. G. Troost, and R. J. Blakely. 2002. "Magnetostratigraphy, Paleomagnetic Correlation, and Deformation of Pleistocene Deposits in the South Central Puget Lowland, Washington." *Journal of Geophysical Research* 107(B4): 2079.

Johnson, S. Y., C. J. Potter, J. J. Miller, J. M. Armentrout, C. Finn, and C. S. Weaver. 1996. "The Southern Whidbey Island Fault: An Active Structure in the Puget Lowland, Washington." *Geological Society of America Bulletin* 108: 334–54.

Polenz, M., H. W. Schasse, and B. B. Peterson. 2006. *Geologic Map of the Freeland and Northern Part of the Hansville 7.5-Minute Quadrangles, Island County, Washington*. Washington Division of Geology and Earth Resources Geologic Map GM-64. www.dnr.wa.gov/Publications /ger_gm64_geol_map_freeland_hansville_24k.pdf.

Stoffel, K. L. 1980. *Stratigraphy of Pre-Vashon Quaternary Sediments Applied to the Evaluation of a Proposed Major Tectonic Structure in Island County, Washington*. Washington Department of Natural Resources, Division of Geology and Earth Resources Open–File 81–292. pubs .usgs.gov/of/1981/0292/report.pdf.

US Geological Survey. Pacific Northwest Geologic Mapping and Urban Hazards. "Current Research—Cascadia Framework and Regional Databases." geomaps.wr.usgs.gov/pacnw/psv1/index.html.

12. Glacial Erratics of the Puget Lowland

Booth, D. B., K. G. Troost, J. G. Clague, and R. B. Waitt. 2003. "The Cordilleran Ice Sheet." *Developments in Quaternary Science* 1: 17–43.

Swanson, T. W., and M. L. Caffee. 2001. "Determination of [36]Cl Production Rates Derived from the Well-Dated Deglaciation Surfaces of Whidbey and Fidalgo Islands, Washington." *Quaternary Research* 56: 366–82.

University of Washington ESS Department. "Washington Glacial Erratics." waglacialerratics.ess.washington.edu/wordpress/. An interactive map and database of significant erratics in Washington.

13. SAVE SNOQUALMIE FALLS!

Dragovich, J. D., et al. 2009. *Geologic Map of the Snoqualmie 7.5-Minute Quadrangle, King County, Washington.* Washington Division of Geology and Earth Resources Geologic Map GM-75.

Easterbrook, D. J. 2003. "Cordilleran Ice Sheet Glaciation of the Puget Lowland and Columbia Plateau and Alpine Glaciation of the North Cascade Range, Washington." In *Quaternary Geology of the United States,* INQUA Field Guide, edited by D. J. Easterbrook, 265–86. Reno, NV: Desert Research Institute.

US Geological Survey. National Water Information System: Web Interface. "Snoqualmie River near Snoqualmie, WA." (Stream gauge located at stop 2.) waterdata.usgs.gov/usa/nwis/uv?site_no=12144500.

14. A TOUR OF DOWNTOWN SEATTLE BUILDING STONE

Courland, R. 2011. *Concrete Planet: The Strange and Fascinating Story of the World's Most Common Man-Made Material.* Amherst, NY: Prometheus Books.

Williams, D. B. 2009. *Stories in Stone: Travels Through Urban Geology.* New York: Walker and Co.

Williams, D. B. 2005. *The Seattle Street-Smart Naturalist: Field Notes from Seattle.* Portland, OR: WestWinds Press.

15. THE FOLDED ROCKS AT BEACH 4

Pazzaglia, F. J., M. T. Brandon, and K. W. Wegmann. 2002. "Fluvial Record of Plate-Boundary Deformation in the Olympic Mountains." In *Field Guide to Geologic Processes in Cascadia,* Oregon Department of Geology and Mineral Industries Special Paper 36, edited by G. W. Moore, 223–56.

16. HURRICANE RIDGE

Babcock, R. S., R. F. Burmester, D. C. Engebretson, A. Warnock, and K. P. Clark. 1992. "A Rifted Margin Origin for the Crescent Basalts and Related Rocks in the Northern Coast Range Volcanic Province, Washington and British Columbia." *Journal of Geophysical Research: Solid Earth* 92: 6799–821.

Carson, B., and S. Babcock. 2009. *Hiking Guide to Washington Geology.* Sandpoint, ID: Keokee Publishing.

Gerstel, W. J., and W. S. Lingley. 2003. *Geologic Map of the Mount Olympus 1:100,000 Quadrangle, Washington.* Washington Division of Geology and Earth Resources Open File Report 2003–4. www.dnr.wa.gov /Publications/ger_ofr2003-4_geol_map_mountolympus_100k.pdf.

Haeussler, P. J., D. C. Bradley, R. E. Wells, and M. L. Miller. 2003. "Life and Death of the Resurrection Plate: Evidence for Its Existence and Subduction in the Northeastern Pacific in Paleocene-Eocene Time." *Geological Society of America Bulletin* 115: 867–80.

Madsen, J. K., D. J. Thorkelson, R. M. Freidman, and D. D. Marshall. 2006. "Cenozoic to Recent Plate Configurations in the Pacific Basin: Ridge Subduction and Slab Window Magmatism in Western North America." *Geosphere* 2: 11–34.

Tabor, R. W., and W. M. Cady. 1978. *The Structure of the Olympic Mountains, Washington: Analysis of a Subduction Zone.* US Geological Survey Professional Paper 1033.

US Geological Society. Geology of the National Parks. "Geology of Olympic National Park: Preface to Online Version." geomaps.wr .usgs.gov/parks/olym/onpreface.html.

17. Dungeness Spit

Evans, O. F. 1942. "The Origin of Spits, Bars, and Related Structures." *Journal of Geology* 50: 846–63.

Schwartz, M. L., P. Fabbri, and R. S. Wallace. 1987. "Geomorphology of Dungeness Spit, Washington, USA." *Journal of Coastal Research* 3: 451–55.

18. Skagit Gorge

Beckey, F. W. 2003. *Range of Glaciers: The Exploration and Survey of the Northern Cascade Range.* Portland, OR: Oregon Historical Society Press.

Brown, E. H. 1989. *Geology of the Skagit Crystalline Core and the Northwest Cascades System, Marblemount Area.* Northwest Geological Society Field Trips in Pacific Northwest Geology. www.nwgs.org/field_ trip_guides/1.%20North%20Cascades.pdf.

Brown, E. H., and J. D. Dragovich. 2003. *Tectonic Elements and Evolution of Northwest Washington.* Washington Division of Geology and Earth Resources Geologic Map GM-52. www.dnr.wa.gov/Publications /ger_gm52_tectonic_evolution_nw_washington.zip.

Brown, E. H., and N. W. Walker. 1993. "A Magma Loading Model for Barrovian Metamorphism in the Southeast Coast Plutonic Complex, British Columbia and Washington." *Geological Society of America Bulletin* 105: 479–500.

Brown, N. 2014. *Geology of the San Juan Islands*. Bellingham, WA: Chuckanut Editions.

McGroder, M. F. 1991. "Reconciliation of Two-Sided Thrusting, Burial Metamorphism, and Diachronous Uplift in the Cascades of Washington and British Columbia." *Geological Society of America Bulletin* 103: 189–209.

Miller, R. B., J. P. Matzel, S. R. Paterson, and H. Stowell. 2003. "Cretaceous to Paleogene Cascades Arc: Structure, Metamorphism, and Timescales of Magmatism, Burial and Exhumation of a Crustal Section." In *Western Cordillera and Adjacent Areas*, Geological Society of America Field Guide 4, edited by T. W. Swanson, 107–35. Boulder, CO: Geological Society of America.

Tabor, R. W., and R. A. Haugerud. 1999. *Geology of the North Cascades: A Mountain Mosaic*. Seattle, WA: The Mountaineers.

US Geological Survey. USGS Geology in the Parks. "North Cascades National Park Geology: World Class and Close to Home." geomaps.wr.usgs.gov/parks/noca/content.html.

19: Washington Pass

Glazner, A. F., J. M. Bartley, D. S. Coleman, W. Gray, and R. Z. Taylor. 2004. "Are Plutons Assembled over Millions of Years by Amalgamation from Small Magma Chambers?" *GSA Today* 14 (4/5): 4–11. www.colorado.edu/geolsci/courses/GEOL5700-PCE/Glazner2004.pdf.

Tabor, R. W., and R. A. Haugerud. 1999. *Geology of the North Cascades: A Mountain Mosaic*. Seattle, WA: The Mountaineers.

20. Larrabee State Park

Mustoe, G. E. 2010. "Biogenic Origin of Coastal Honeycomb Weathering." *Earth Surface Processes and Landforms* 35: 424–34.

Mustoe, G. E. 1982. "The Origin of Honeycomb Weathering." *Geological Society of America Bulletin* 93: 108–15. www.tafoni.com/References_files/Origin%20of%20Honeycomb%20Weathering.pdf.

Rodriguez Navarro, C. 1998. "Evidence of Honeycomb Weathering on Mars." *Geophysical Research Letters* 25: 3249–52.

Tafoni.com. "Tafoni." www.tafoni.com/Welcome.html.

21. *Diatryma*

Andors, A. V. 1995. "*Diatryma* among the Dinosaurs." *Natural History* 104 (6): 68–71.

Mustoe, G. E., and W. L. Gannaway. 1997. "Paleogeography and Paleontology of the Early Tertiary Chuckanut Formation, Northwest Washington." *Washington Geology* 25 (3): 3–18. www.dnr.wa.gov/Publications/ger_washington_geology_1997_v25_no3.pdf.

Mustoe, G. E., D. S. Tucker, and K. L. Kemplin. 2012. "Giant Eocene Bird Footprints from Northwest Washington, USA." *Palaeontology* 55: 1293–305.

Patterson, J., and M. G. Lockley. 2004. "A Probable *Diatryma* Track from the Eocene of Washington: An Intriguing Case of Controversy and Skepticism." *Ichnos* 11: 341–47.

Witmer, L. M., and K. D. Rose. 1991. "Biomechanics of the Jaw Apparatus of the Gigantic Eocene Bird *Diatryma*: Implications for Diet and Mode of Life." *Paleobiology* 17: 95–120.

22. ARTIST POINT AND HEATHER MEADOWS

Aydin, A., and J. M. DeGraff. 1988. "Evolution of Polygonal Fracture Patterns in Lava Flows." *Science* 239: 471–76.

Hildreth, W. 1996. "Kulshan Caldera: A Quaternary Subglacial Caldera in the North Cascades, Washington." *Geological Society of America Bulletin* 108: 786–93.

Hildreth, W., J. Fierstein, and M. Lanphere. 2003. "Eruptive History and Geochronology of the Mount Baker Volcanic Field, Washington." *Geological Society of America Bulletin* 115: 729–64.

Tabor, R. W., and R. A. Haugerud. 1999. *Geology of the North Cascades: A Mountain Mosaic*. Seattle, WA: The Mountaineers.

INDEX

Page numbers in boldface refer to photos, illustrations, or information in captions.

DAVE TUCKER lives in Bellingham, Washington. He has a master's degree in geology and is a research associate in the geology department at Western Washington University. He is also a director of the Mount Baker Volcano Research Center, an all-volunteer nonprofit organization that raises funds to support research at the active volcano and to educate the public about the mountain's volcanic hazards (learn more at http://mbvrc.wordpress.com/).

Tucker has been mapping Baker's geology since the mid-1990s, in particular the distribution of volcanic ash deposits. He leads public field trips and gives presentations about the geology of northwest Washington, and he is the author of a popular blog, *Northwest Geology Field Trips* (http://nwgeology.wordpress.com/).